Listed

Listed

Dispatches from America's Endangered Species Act

Joe Roman

Harvard University Press • Cambridge, Massachusetts • London, England • 2011

FOR LAURA AND NGAN

Library of Congress Cataloging-in-Publication Data

Roman, Joe.
Listed : dispatches from America's Endangered Species Act / Joe Roman.
        p.   cm.
Includes bibliographical references and index.
ISBN 978-0-674-04751-8 (alk. paper)
1. Endangered species—United States.   2. Nature conservation—
Government policy—United States—History.   3. Wildlife conservation—
Government policy—United States—History.   4. Environmental
protection—Economic aspects—United States.   5. Environmental policy—
United States—History.   6. Endangered species—Law and legislation—Political
aspects—United States.   7. Environmental law—Social aspects—United
States—History.   8. United States. Endangered Species Act of 1973.   I. Title.

QH76.R64 2011
333.95'22160973—dc22   2010047785

Contents

All great crime stories are stories of redemption.

—Raymond Chandler

"No woodpeckers," read the real estate listing. "Lovely waterfront lot on Shadow Lake. . . . Septic evaluation being applied for." "Oversized lot," read another, "not in woodpecker-affected area." When news had spread around Boiling Spring Lakes in 2006 that the Fish and Wildlife Service was concerned about the effects of rapid development on the red-cockaded woodpecker, residents rushed to City Hall, clamoring for permits to fell the mature pines on their property. Within weeks, thousands of mature longleaf pines in this coastal North Carolina town had been reduced to bark and sawdust. Newspapers reported that the run on chainsaws had emptied store shelves around the county, with landowners reasoning that if the woodpecker's habitat was destroyed before restrictions took hold, they wouldn't have to worry about building permits. "It's ruined the beauty of our city," Mayor Joan Kinney told the Associated Press.[1]

Two years after the chainsaw rush, I stood in the middle of a deserted street. It was still a chilling chain of events.

The Endangered Species Act has been accused of encouraging property owners to clear their land of native habitat. The *Developer's Guide to Endangered Species Regulations,* published by the National Association of Home Builders in 1996, acknowledged that "the highest level of assurance that a property owner will not face an ESA issue is to maintain the property in a condition such that protected species cannot occupy the property. . . . This is referred to as a 'scorched earth' technique."[2] If a species such as the red-cockaded woodpecker liked longleaf, chop them down. No pines, no problems. Some landowners shortened the time between timber harvests to prevent the protected birds from nesting on their land, thus avoiding restrictions. The number of woodpecker groups in the North Carolina Sandhills declined from 590 in the 1980s to 371 in 1990, in part because of preemptive harvests.[3]

Similar reactions could be found through much of the United States. In 1999, an Alabama congressman had warned that listing the Alabama sturgeon, a fish so rare that only eight had been captured in the previous decade, would "threaten our communities, our businesses and our homes and our daily lives." Another congressman testified that more than 115,000 jobs were "in the crosshairs of the Fish & Wildlife Service" because of the listing.[4] But after ten years, the impact has been only minimal: most of the streams had already been protected for freshwater mussels anyway—Alabama has 48 protected species of pearly mussels. Though habitat has been protected on more than 320 miles of waterways in the state, it's unclear whether any sturgeon are left to use it. Only two have been captured and released since listing.

In Wyoming, when I stopped at a bar on Chief Joseph Highway, the bartender told me I didn't want his view on endangered wolves. "I'd like to kill every one of them. They're nice to look at, but they have their place."

Where's that?

"Yellowstone, if they could keep them there. Put a few on the Mall in Washington, a few in Central Park, a few in Boston Common."

"Get 'em closer, so people won't spend so much on fuel," said one of the customers, handgun on his belt.

Some of *those* people argued that wolves should be free to roam anywhere without persecution even if they attacked valued livestock. Ed Bangs, who helped reestablish wolves to the Northern Rockies, called them "the hundred-pound club that people use to bash each other over the head with."

The Endangered Species Act is under attack. It has long been under attack. Even more than other environmental achievements of the 1970s—the Clean Air Act, the Clean Water Act, the Pesticide Control Act—the Endangered Species Act draws anger: It doesn't work. It costs too much. It puts people out of work. It puts animals before humans.

But our understanding of ecology, epidemiology, and economics has changed since the Endangered Species Act was signed. In 1973, traditional economics, which looked to self-interest and endless economic growth as the predominant drivers of the economy, was the only game in town. Since then, ecological economics, which places the economy within the global ecosystem, has restored a view closer to that held during the Enlightenment: Nature could be beneficent. Wetlands and reefs protect communities from storms. Wild habitats, including those set aside for rare and endangered species, provide clean air and fresh water and regulate the climate—all for free. The field of epidemiology began to take a few ecologists onboard. Diseases such as HIV/AIDS, malaria, Lyme, and hantavirus had emerged or reemerged as we altered native ecosystems or released invasive species. In contrast to the long-held view that human health was largely an internal matter between a few domesticated species and us, scientists have shown that protecting and restoring native ecosystems could be critical to controlling new diseases.

In the 1970s, climate change had yet to emerge as an environmental issue. Now, the threat of global warming has put us all in the same ark: a 4°C rise over preindustrial levels, possible in the next 50 years, could swamp low-lying coasts, threaten freshwater supply, and doom more than a third of Earth's animal and plant species to extinction.[5] Was the recognition that we were all in this together—and that by preserving species and their habitat we could slow these changes down, if not put a stop to them—a game changer?

As I started this book, I chose a few cases where there appeared to be a clear conflict between conservation and economics: the red-cockaded woodpecker that knocked down a city, a pair of mussel species in Apalachicola that sucked another one dry, the wolves that outhunted elk hunters. Were such species actually being protected? How was this protection—or extinction—affecting people? Almost 40 years after the Endangered Species Act became law, I wanted to see if biodiversity protection worked.

1

A small, ghostly fish looked out from a pool of ink, beside it a half-completed dam. This is my earliest memory of the Endangered Species Act: a front-page article in the *New York Daily News* circa 1977. CBS News reported it as a mismatched prizefight: "In this corner, weighing 156,000 tons, the source when completed . . . of enough electricity to power a small town, the Tennessee Valley Authority's mighty Tellico Dam. And in this corner, weighing less than one half ounce, measuring two-and-one-half inches, the snail darter, a contender with high hopes. . . . The snail darter figures to punch a hole in the dam."

The fish, which didn't even have a scientific name when the Endangered Species Act was passed in 1973, was slender, with four dark saddles and showy purple caudal fins if you caught it in a certain light. Its lateral fins were like upside-down wings; rather than lifting the fish up, they kept it down, just above the river bottom, where it fed. It was no mighty bull trout, no Chinook salmon. Cecil Andrus, secretary of the Interior during the battle over the darter, complained that it was the only fish story he knew where the fish kept getting smaller each time the tale was told. But the snail darter had a spectacular home, one of the last open stretches of running water in eastern Tennessee, well known to anglers for its trout fishing. With clear-water riffles and gravel runs, the Little Tennessee River wound through some of the finest soils in the state. In the early 1970s, earthmovers from the Tennessee Valley Authority were getting ready to flatten it all.

The Authority had been building dams throughout the Tennessee Valley since 1933. Perhaps no region suffered more during the Great Depression: approximately one in three people had malaria, crops had failed, and most of the timber had been cleared. To bring work and electricity to the area, Congress had created a public corporation, the Tennessee Valley Authority, which quickly drafted a list of 60 potential dam sites in the valley. Each of the TVA's impoundments would be multipurpose: to generate electricity, increase navigation, and control floods—goals that previously had been seen as conflicting. It was the first public-works project designed to control and develop the water of an entire river system.

The Authority had the power of eminent domain and at first met little organized resistance. By 1948, it had built three dozen dams on the prime river sites. But the money kept coming in from Congress, and the Authority continued to work its way down the list, damming so many rivers and tributaries that Tennessee soon had more shoreline than all the Great Lakes combined.

Construction at the site on the bottom of the list, a relatively small one on the Little Tennessee River, didn't begin until 1967. The Authority promoted the 129-foot Tellico Dam as the source of cheap electricity, but in truth no direct power would be generated. A canal to the nearby Fort Loudon plant would boost energy production by less than a tenth of a percent. The twisting Little Tennessee had no upstream ports, so the structure, which would span more than half a mile across the river, would do little for navigation. With 30 dams in the area already, there was little need for further flood control. Still, every time the downstream city of Chattanooga had an overflow, the TVA lamented that the Tellico Dam would have prevented it.

Since neither of the conventional benefits of power generation and reduced flooding could justify the estimated $40 million in construction costs, the Authority's main economic justification came down to the projected industrial development that would occupy five thousand acres of land along the new lakeshore. (Before he had signed the bill creating the TVA, Roosevelt had insisted on including economic development as one of the Authority's key missions.) To fulfill the promise of twenty-five thousand jobs, a model city of fifty thousand, called Timberlake, was planned.[1] The Boeing Corporation had expressed interest in the project. And why not? With the power of eminent domain, the TVA could condemn farmland at a low price (an average of only $356 an acre), then turn around and sell the new waterfront property at any rate it could fetch from private developers. Flipping real estate was at the heart of the Authority's economic analysis.

Not everyone saw this as a good deal. Local farmers, fishermen, and several environmental groups, led by the Environmental Defense Fund, insisted that too much would be lost to justify the dam's completion. The EDF filed for an injunction to stop the project until an Environmental Impact Statement was completed to study the environmental costs of the project. The newly enacted National Environmental Policy Act required that all federal agencies prepare such a statement. The TVA refused—after all the concrete portion of the Tellico Dam had already been poured and only the construction of the earthen embankment remained to close off the river. The EDF won the case, and under court order, the Authority drafted an EIS. But the statement was seen as perfunctory, failing to recognize the biological importance of the river. The EDF prepared to go to court again.

David Etnier, an ichthyologist at the University of Tennessee, was expecting to be called as an expert witness. He thought it might help his testimony to go down to the river before the trial. A friend had told him about an area where they could set out decoys and duck hunt, but it had been so foggy they could barely find a place to put in. "We saw one grebe," he told me. "But the river was beautiful, and the landowner was really pleasant." Since the landowner had told them to come back anytime, in August 1973, Etnier and a graduate student returned with face masks and snorkels, heading for the end of the river, the warmest place, where the probability of finding unusual native species was highest.

"We jumped in the river and started looking for fishes," Etnier said. "The first fish I saw I thought was a sculpin, which are common in the river. But it looked a little too slender. Sculpins have great big heads. This one didn't seem to have nearly a big enough head. The visibility was pretty good, but I was still a little confused by it. So I poked at it and it swam ahead about two feet and settled down to the bottom. I cradled my two hands around it and was able to just pick it out of the water. I could tell that I had a darter that nobody had ever seen before."

Etnier, who knew the darters of North America, having collected thousands of fish in Tennessee, walked over to a gravel island with his student. They both looked at it. "Wow, wow, this is incredible!" he said. It closely resembled a species restricted to the Ozarks in Arkansas. "Instead of finding one of these rare and endangered species, I had found a brand new one."

The thrill of scientific discovery is a highlight of any zoologist's career, but to find a unique and undoubtedly rare species on the eve of a court case, a fish that might help him save his favorite trout river—this was big. They returned to get additional specimens for description, keeping the fish alive overnight. "The next morning the bottom of the cooler was littered with snail shells." Etnier had a common name for his new species: the snail darter.

The plaintiffs in the EDF case called nine witnesses (including economists, archaeologists, and biologists) to challenge the TVA's impact statement. One botanist described the TVA's species list as "grossly inadequate, incomplete." An agricultural expert testified that the farmlands had been undervalued. The lawyers knew about Etnier's discovery but never called him. "Testifying is not all that much fun," Etnier told me. "It didn't bother me a bit, but," he paused, "we lost." Judge Robert Love Taylor conceded that "more detailed studies could have been performed," but deferred to the TVA, ruling that their economic and environmental considerations were adequate. He acknowledged that the $35 million already spent on building the dam played a part in his deliberations as he dismissed the suit.[2]

A few months after the court ruling, Congress passed the Endangered Species Act. Those who wanted to protect the river and its farmlands would have one more chance.

Back at the University of Tennessee, Etnier's work caught the attention of Hank Hill, a student at the law school who had heard the story over a few beers. He thought the snail darter might make an interesting paper for his environmental law course. Zygmunt Plater, the course instructor, became convinced Etnier's darter could be protected under the new law. In January 1975, he and Hill petitioned the Fish and Wildlife Service to list the new fish as endangered.

"We all knew that the darter would have a better chance of being listed if it was formally described," Etnier told me, "rather than just some ichthyologist saying that they had an undescribed species down there." Across the span of 35 years, he recalled having published his paper in a peer-reviewed journal a few months after the passage of the act in 1973. My reading of the case suggested it had taken a bit longer. He looked it up on his CV. "Nineteen seventy-six. Wow, that did take a while. I thought I was getting to be a terrible procrastinator, but I guess I was pretty slow in those days, too."

To be fair, it takes time to describe a new species. You need to count scales down the side of the fish, then along the lateral line and around the tail, tallying up the spines and rays in the fins and the vertebrae—and do this for about a dozen individuals. You have to examine internal organs such as sensory canals and gills and describe the pigments, stripes, and blotches in all their variations. You tease apart the difference between males and females (the latter typically much less colorful). You store the holotype (the official record for the species) and the related paratypes, then review all the literature to assure that your new species won't later be lumped by lumpers—who viewed closely related organisms as incipient species that could interbreed, questioning their unique status—or split by splitters, who saw emergent species everywhere. Then you do the math: the distribution of these characteristics has to fall outside the measurements for related species—in the case of the snail darter, populations from three river systems in the Ozarks.[3] (It didn't help Etnier that the taxonomic status of that species was in flux.) You submit the paper for review, hoping to get it accepted and eventually published. Oh, and then there's the time required for teaching and maintaining a lab. That this took a couple of years didn't seem outside the norm—unless you had the TVA breathing down your neck.

While Etnier worked on his description, the director of the Fish and Wildlife Service issued a proposed rule to list the snail darter, inviting public comments. The TVA was undeterred. In the House of Representatives, the Appropriations Committee recommended that funds be secured for Tellico and the dam be completed without delay. The gates could be closed before the fish was described. It was the saddest of storylines of the thousands of extinction tragedies: a species disappears just as it is discovered.

In October 1975, Plater and colleagues sent a letter to the TVA and Department of the Interior to stop further construction of the dam until the listing decision was made. He and Hill asked the secretary of the Interior to use the emergency provisions of the act, arguing that the imminent completion of the Tellico could wipe out the entire species. Jim Williams, an ichthyologist from Mississippi who had recently been hired by the Service, was convinced that the petition had merit, but his boss, Keith Schreiner, "was scared shitless of this whole thing and had been trying to block it," Plater told me. Williams said that Schreiner's "heart was in the right place," but he was meticulous and under pressure from Capitol Hill and other quarters of the Interior Department. Even Schreiner finally saw that the case seemed irrefutable.

The fish was listed in the *Federal Register* the following week. A couple of months later, Plater and Hill filed for a permanent injunction forbidding the completion of the dam, employing Section 11, the Endangered Species Act's citizen-suit provision. A "proud, very American innovation," according to Plater, the provision gave the darter (or at least its watchdogs) statutory standing to enforce the law in court against powerful adversaries such as the TVA, part of a shift in American jurisprudence.[4] In a famous decision in *Sierra Club v. Morton* in 1972, the Supreme Court had ruled that the Sierra Club lacked standing to stop a ski resort in the Sequoia National Forest because it hadn't shown injury to any of its members. The dissent, straight out of Aldo Leopold, one of the intellectual founders of the modern conservation movement, came from Justice William O. Douglas: "Before these priceless bits of Americana (such as a valley, an alpine meadow, a river, or a lake) are forever lost or are so transformed as to be reduced to the eventual rubble of our urban environment, the voice of the existing beneficiaries of these environmental wonders should be heard.

"Perhaps they will not win. Perhaps the bulldozers of 'progress' will plow under all the aesthetic wonders of this beautiful land. That is not the present question. The sole question is, who has standing to be heard?"[5]

Plater, Hill, and the farmers of the Little Tennessee Valley had plenty of standing, but they were short of funds. The protestors were supported by the sale of T-shirts—a play on the *Jaws* poster, with "TVA" spelled out in shark teeth—and fueled on potlucks. The director of the Fish and Wildlife Service notified the Authority that a listing would act as "a mandatory prohibition" of any federal action jeopardizing the darter's continued existence.[6] Construction of the dam would have to stop. So all of this came from Hill's law-school paper? Plater laughed when I asked. "Oh, I don't think the term paper was ever turned in."

The TVA's chairman rushed up to DC, testifying before Congress that the price on the Tellico had already gone up because "extensive litigation" increased costs and lengthened delays. This would become a familiar refrain from opponents of the Endangered Species Act and later from the biologists and administrators in charge of the listing process itself.

Much more was at risk than the darter. The Little Tennessee, a favorite among fishermen, had been considered the finest large trout river in the Southeast even before the TVA and the Army Corps of Engineers had flattened the competition.[7] Some of the most valuable farmlands in eastern Tennessee and areas of great recreational, historical, and archaeological value—including a fort from the French and Indian War—would be inundated by the dam. Sites sacred to the Cherokee lay close to the Little Tennessee, including the city of Echota, described as the group's Jerusalem. Upstream, the Chilhowee Dam, built by Alcoa in 1957, had already submerged several sites; the Tellico would drown the rest. One bumper sticker read: "Don't Bury My Heart at the Little T."

The TVA condemned more than thirty-five thousand acres, including entire farms that had been in cultivation for generations. Only a third of this would be flooded—but the TVA wanted all of it. It could buy low—the price of condemnation did not include the value of 25 feet of rich topsoil or of the farming community that had grown on it—and resell the surplus at a higher price. Out of the 340 families that would lose their land by eminent domain, many took the government's offer. But others refused to leave their homes, joining sport fishermen, historic preservationists, and the Eastern Band of the Cherokee Indians in opposing the project.

And it was all coming down to the darter. The TVA pounced on Etnier's delay in publishing his new species. In a letter to the chief of the Office of Endangered Species in the spring of 1975, the Authority's director of Forestry, Fisheries, and Wildlife Development challenged Etnier's description of "the so-called 'snail darter' which allegedly exists in the Little Tennessee River."[8] The TVA refuted the species-level designation of the darter and its narrow range (between miles five and seventeen on the Little Tennessee), employing the expertise of renowned ichthyologist Edward Raney, a retired professor from Cornell. Raney flew over the area, but didn't bother to seine it. He pronounced that there were plenty of rivers where the darter could exist. The Authority's biologists would spend several years and close to a million dollars searching for it. "They looked in all these itty bitty creeks," Etnier told me, "rediscovering the banded sculpin about four or five hundred times in Tennessee." A sculpin. If ichthyologists have a trash fish (and as far as I know they don't), the common sculpin would be it.

Etnier had his own ideas about darter habitat, but the TVA's lawyers told their biologists to stay away from him. "My first choice," Etnier said, "would have been to go to the Hiwassee," a river that drains into the main branch of the Tennessee several miles south of the Little T. "The reason for that is there's a caddis fly that I'm familiar with, with a very, very restricted distribution. It's only in three rivers: one is in north Georgia, the Little Tennessee River, and the Lower Hiwassee. Anything that's that restricted is probably a complete habitat specialist." Just like the snail darter. No one at the TVA ever asked Etnier what he thought, except when they challenged him in court.

The snail darter was listed as endangered in November 1975. That same year, Boeing abandoned its Timberlake project. A few months after the listing, Judge Taylor reviewed the request for a permanent injunction against the Tellico. He acknowledged that the Tellico would probably eliminate "almost all of the known population of snail darters," but again he refused to stop the dam. In his view, the Act's problems had not arisen until long after the TVA had begun work on the Tellico, and the Authority had made a "good faith effort" to conserve the darter, supporting several research projects, including that of Etnier's grad student.[9] The law, Taylor noted, should not be enforced when only a little fish "or some red-eyed cricket" would be extinguished.

Both chambers of Congress recommended appropriating $9.7 million. A Senate subcommittee declared that it did not view the Endangered Species Act as prohibiting the completion of the dam; it should be finished "as promptly as possible." Land-clearing teams worked round the clock. Night shifts cut trees under portable lights. Their goal, according to an inside source, was to make sure that by the time Plater stood up to argue his appeal, there wouldn't be a tree left standing in the valley that would be drowned by the reservoir.[10]

There were still a few of them standing when the Sixth Circuit reversed Taylor's decision, without dissent. "Whether a dam is 50 percent or 90 percent completed is irrelevant in calculating the social and scientific costs attributable to the disappearance of a unique form of life," Judge Anthony Celebrezze said in rejecting the TVA's recurrent argument that too much had already been invested to stop. Congressional appropriations could not override the law. "Courts are ill-equipped to calculate how many dollars must be invested before the value of a dam exceeds that of an endangered species. Our responsibility is merely to preserve the status quo where endangered species are threatened." As there was no exemption in the act for Tellico, the Sixth Court ruled that the lower court had "abused its discretion" in letting the TVA continue work on the dam.[11]

Zygmunt Plater was at Wayne State University in Detroit when he got the news, having been fired from the University of Tennessee soon after he presented the snail darter case. "TVA told the law school off the record that as long as I was there, there wouldn't be any more TVA scholarships for students. Then they got a complaint filed against me for barratry, trying to drum up legal business. It was messy, it was dirty. I had tremendous publications and tremendous teaching evaluations, but as one of the senior guys told me, I did not understand the moderation required of Tennessee law professors." Etnier told me Plater's colleagues had denied that the snail darter had anything to do with their decision.

I asked Plater if the commitment had been worth it. "It was one of those things that we couldn't not do." His family had lost its land—twenty thousand acres in Poland—to the Nazis, and then to the Communists. He had spent two years in the Peace Corps in Africa. "We had all the necessary facts," he said. "We had the law and common sense on our side. If we didn't have enough faith in the American legal system to try it, then what are we going to think about ourselves or tell our grandchildren?"

It was pure celebration when they got word from the Sixth Circuit. "There was plenty of snow on the ground," said Etnier, who was in Ann Arbor. "We heated up the sauna, rolled around in the snow, went back to the sauna. We probably had a few too many beers." Ah, the seventies.

Its work halted once again, the TVA worked the halls of Congress to secure more appropriations. The petition it submitted to the Supreme Court would lead, in the coming year, to the highest court in the land issuing the most important decision in the history of American endangered species conservation.

2

On March 11, 1967, the Department of the Interior released an official list of fish and wildlife threatened with extinction in the United States, the first wide-scale effort to address the extermination of native species. Seventy-eight animals, all vertebrates, made the short list. There were 14 mammals, including the black-footed ferret and the Florida panther. There were 36 birds, among them icons such as the southern bald eagle, the whooping crane, and the ivory-billed woodpecker (not seen since 1944). There were 3 reptiles (the American alligator, the blunt-nosed leopard lizard, and the San Francisco garter snake), 2 salamanders, and a toad. The 22 fish species spent all or most of their lives in freshwater.[1] This "class" wasn't the only list of threatened wildlife, but it was the first with legal backing and it caught the public's imagination.

The Endangered Species Preservation Act was remarkably frank as to the cause of the threats: "one of the unfortunate consequences of growth and development in the United States has been the extermination of some fish and wildlife."[2] Congress instructed the Interior Department to seek the advice of ornithologists, ichthyologists, ecologists, herpetologists, and mammalogists and to acquire lands to conserve these species—but not to spend more than $15 million a year making purchases. All of the species would be protected "insofar as is practicable." In introducing the list, Secretary of the Interior Stewart Udall didn't hedge: "I find after consulting with the States, interested organizations, and individual scientists that the following listed native fish and wildlife are threatened with extinction." He continued, "We hope that with man's aid and understanding, the long list of threatened species can be narrowed." But the numbers have been growing ever since.

Perhaps the first zoologist to compile such a list, comprised of at-risk mammals and birds, was Joseph A. Allen, a curator at Harvard's Museum of Comparative Zoology. His expeditions in the West—he was among the first to warn of the approaching extinction of the bison—and his careful scrutiny of the accounts of early explorers and naturalists resulted in a series of academic and popular articles that came out in 1876, just as the United States was celebrating its centennial. Allen was painfully shy, and his groundbreaking work had been largely overlooked until it was rediscovered recently, thanks to the research of historian Mark Barrow.[3]

The first systematic attempt to list all the species "threatened with early extermination" was William T. Hornaday's *Our Vanishing Wild Life,* published in 1913. An eminent zoologist, renowned taxidermist, and anything but shy, Hornaday contacted wildlife managers in every state and Canadian province, gathering their opinions on threatened species. He was a natural list maker: "Name some of the species?" he prompted himself. "Certainly, with all the pleasure in life." In California: "the band-tailed pigeon, the white-tailed kite, the sharp-tailed grouse, the sage grouse, the mountain sheep, prong-horned antelope, California mule deer, and ducks and geese too numerous to mention." He admitted that it would take years of work to get to figures that were "absolutely exact," but his was the first continental survey of the status of birds (and a few mammals). The list included 23 North American avian species and 4 mammals as "candidates for oblivion" and an epitaph "Sacred to the Memory" of 11 North American birds, including the great auk, the Labrador duck, and the passenger pigeon, which were "Exterminated by Civilized Man 1840–1910." Some species on his list would eventually bounce back, such as the willet and the dowitcher; others (such as the heath hen and the Caribbean monk seal) would later disappear.[4]

Hornaday saw one ray of hope—the American bison. Early in his career, he'd helped establish the National Zoological Park in Washington, DC, negotiating the 166-acre site and testifying before Congress to obtain funding. (Even earlier, he had shot several dozen buffalo in Montana, including a massive old bull, to display at the Smithsonian. It's still there.) He believed the park could help save the nation's "great game animals," but abandoned the project over con- . flicts with the head of the museum. By 1913, after Hornaday assumed directorship of the New York Zoological Society, the bison's future, if greatly diminished, was reasonably secure: "The efforts of man to atone for the great bison slaughter by preserving the species from extinction have been crowned with success. Two governments and two thousand individuals have shared this task,—solely for sentimental reasons. In these facts we find reason to hope and believe that other efforts now being made to save other species from annihilation will be equally successful."[5] Bison from the New York zoo would eventually help restock the West.

Even as he watched game disappear, Hornaday believed in the value of predators. A real sportsman, he insisted, opposed the extermination of the grizzly bear: "A Rocky Mountain without a grizzly upon it, or at least a bear of some kind, is only half a mountain,— commonplace and tame. Put one two-year-old grizzly cub upon it, and presto! every cubic yard of its local atmosphere reeks with romantic uncertainty and fearsome thrills."[6] Aldo Leopold, who would later run with this mountain idea, was on leave from the Forest Service when he read Hornaday's book soon after it was released. It helped persuade him that man was responsible for protecting threatened game species, and he later praised Hornaday's work as "one of the milestones in moral evolution."[7] Hornaday made certain that this book, funded by his own wildlife protection fund, would not go unnoticed; he sent a copy to every member of Congress.[8]

It would be several decades before the first official list was released. In 1942, the American Committee for International Wild Life Protection published *Extinct and Vanishing Mammals of the*

Western Hemisphere. Six years in the making and more than six hundred pages long, it was a comprehensive review, from the recovery of bones of the extinct Puerto Rican ground sloth in limestone caves to endemic rice rats of the Galápagos and the walrus-hunting habits of the Eskimos. The lead author, Glover Allen, died of heart failure before the book was released. Having spent years studying species in Australia, Africa, and North and South America while working at the Museum of Comparative Zoology, Allen might have seen the work as "the clearing of a new trail into a virgin forest."[9] The committee hoped it would encourage international cooperation to preserve vanishing wildlife. Among the scores of recommendations, it called for a Pacific walrus reserve in Alaska to protect the declining oceanic mammals and their native hunters from the waste of commercial sealing.

During that same year, a group of biologists from the National Park Service and Fish and Wildlife Service published a book for the general public. *Fading Trails: The Story of Endangered American Wildlife* had chapters on 25 "vanishing species," as they called them, and one catchall, "rare furbearers." Most of the species covered (such as the grizzly bear, the condor, the sea otter, and the whooping crane) were included on later lists. The only invertebrate, the lobster, stands out as the single species that continues to have a strong commercial market. When the book was written, the fur trade was thriving and the states had legal control over trapping. "Perhaps the only place in the United States where our grandchildren will ever see a marten, a fisher, or a wolverine will be in a museum," the authors wrote.[10] I suppose I could be one of those grandchildren. Efforts to control the fur trade made in response to the grave prognosis for these early champions have met with some success. I've been fortunate to see fishers during my hikes in the East, martens have returned to the nearby Adirondacks, and wolverines persist on the tundra and taiga, though they're still at risk.

Vanishing was the common term for rare species until the 1960s, when *endangered* or *threatened* came to be used, a shift hammered home with the passage of endangered species legislation. In addition to *Fading Trails,* one of the earliest appearances of *endangered* was in the paper "Conserving Endangered Wildlife Species," published in 1943 by Hartley H. T. Jackson, who was in charge of the Biological Surveys at the Department of the Interior.[11] One of the biologists working on the later lists told me that while *vanishing* had a certain emotive value that could appeal to the public, it wasn't very precise. The words *threatened* and *endangered* were less ambiguous, and they implied that you could take the dangers away and restore a species: *vanishing* felt as if the situation had been removed from our control. The new term took on a life of its own. The American middle class is perpetually an "endangered species"; so, too, are Democrats (early 2000s); Republicans (later 2000s); Palestinians and Israelis (depending on whom you read); Vulcans (Spock, 2009); dentists; dairy farmers; receptionists; muscle cars; economic growth; single, good men *(The Onion);* and writers, along with linear, deep-focus reading itself *(The Wall Street Journal).*

What is now the world's most comprehensive list of threatened species began rather modestly, with a draft containing 27 gravely endangered ones (14 mammals and 13 birds) chosen by delegates at the temporary headquarters of the United Nations in Lake Success, New York, in 1949. The International Union for the Protection of Nature had been founded a year earlier as a loose affiliation of governments, environmental organizations, and scientists. Harold Coolidge, an American mammalogist, worked to strengthen the network and secure funding. From the start, its primary concern was the threat of extinction. Julian Huxley, who as the director-general of UNESCO had been influential in getting the union started, wrote to his brother Aldous about his concern for the "cosmic capital on which we live."[12]

In 1954, a young ecologist, Lee Merriam Talbot, was just out of the Marine Corps when he landed his dream job: he was sent to survey the status of several large mammals in Egypt, Syria, India, the Himalayas, Nepal, Sumatra, and Java—areas that hadn't been studied since well before the war, if at all—and spread the IUPN's message. "The issues often revolved around loss of habitat as well as over-exploitation," he told me recently. "I walked down the coast of Sumatra looking for tigers and the Sumatran rhino, which I found. In many villages they told me that I was the first white person they had seen since the Dutch in the thirties." He published journal and magazine articles about the surveys and would eventually get his PhD—his "union card," as he described it—become director of the union, and play a crucial role in the drafting of the Endangered Species Act. Soon after Talbot's trip, India's prime minister, Jawaharlal Nehru, visited the Gir Forest and declared it a national park.

In 1956, the union traded "Protection" for the softer "Conservation" to become the IUCN, in part to appeal to the political concerns of countries that wanted to use their natural resources to build economies, not just to preserve "nature" or wilderness. Leopold had earlier promoted the word *conservation* because to him it meant "a positive exercise of skill and insight," not abstinence or caution.[13] The Europeans, Talbot told me, initially resisted the change because they associated the word with making jams. They relented, and the IUCN, or World Conservation Union, continued compiling lists. As they became more comprehensive, these Red Lists—the color was chosen to emphasize the imminent danger of extinction—grew into databases: Red Data Books. The first volumes on mammals and birds were printed in 1966, followed by reptiles and amphibians, then flowering plants and fishes. The Red List is now so authoritative, its name is trademarked.

These lists were essential in raising awareness of endangered species, but they were still only *lists,* and the narratives had to be teased out from the numbers and Latin binomials. In 1962, a bestseller about a molecule changed the landscape. In 1958, Rachel Carson pitched a story to *The New Yorker*'s E. B. White, suggesting that he write about the effects of pesticides on birds. He turned her idea down but suggested she tackle the subject herself. William Shawn, the magazine's editor, saw the potential and assigned her fifty thousand words, a book-length series of articles. Carson's was an apocalyptic vision, a "world where the enchanted forest of the fairy tales has become the poisonous forest in which an insect that chews a leaf or sucks the sap of a plant is doomed. It is a world where a flea bites a dog, and dies because the dog's blood has been made poisonous, where an insect may die from vapors emanating from a plant it has never touched, where a bee may carry poisonous nectar back to its hive and presently produce poisonous honey."[14] The world of insecticides, according to Carson, "surpasses the imaginings of the brothers Grimm."[15] It was a dark fairy tale, lit by the fiery memories of Hiroshima, Dresden, and Auschwitz. "Under the philosophy that now seems to guide our destinies, nothing must get in the way of the man with the spray gun."[16] White predicted *Silent Spring* would be as influential as *Uncle Tom's Cabin*.[17]

The book caught heat. A spokesman from the chemical industry accused Carson of making gross distortions, "completely unsupported by scientific, experimental evidence."[18] The protests, and the book's appearance as a serial in that most urban of magazines, *The New Yorker,* in 1962, assured it an audience. Carson may have been a slow writer, but her timing proved impeccable. As her book was released, radioactive isotopes from atomic bomb tests in the West and South Pacific had been found in rain falling over New York City—and in the red blood cells of the city's children, thanks to the milk they drank. Carson defined the contamination from pesticides and the fallout from atomic war as "the central problem of our age."[19] Then the thalidomide scandal broke: babies with limbs malformed by a medication their mothers had been prescribed for morning sickness. *Silent Spring* caught fire, selling more than 40,000 advance copies and 150,000 through the Book of the Month Club. President Kennedy read it and appointed a scientific committee to study the toxicity of pesticides.

By focusing on the decline of birds, Carson helped bring species extinctions to the forefront of the nascent environmental movement.

In July 1964, the Department of the Interior published its first official list of extinct and endangered fish and wildlife, the predecessor to the Class of '67, the first group to have legal protections. The list, informally compiled by a team of nine biologists, was introduced by Ralph Yarborough, a liberal Texas democrat who was the only Southern senator to vote for the Civil Rights Act. Thirty-nine species that were known to have gone extinct since the United States was formed were entered into the Congressional Record, including the Oahu thrush (around 1825), Steller's sea cow (1854), Labrador duck (1875), Texas grizzly (1890), sea mink (1890), plains wolf (1895), passenger pigeon (1914), heath hen (1932), and Laysan Island rail (1944).[20]

A list of 59 endangered vertebrate species and subspecies followed. Of the birds, the California condor and the whooping crane have become celebrities. The more modest Nihoa finch remains in danger, as do several other Hawaiian birds. The green turtle, American alligator, American crocodile, and the cutthroat trout were on it (after centuries of overharvesting). The polar bear and the walrus were short-timers; they weren't in the Class of '67, but they would become candidates for protection again in the next century under circumstances that would have been entirely foreign to the founding list-makers. The black-footed ferret, thought to have disappeared entirely, was resurrected much later by a rancher's dog. The government now had its list of victims—although it didn't say much about the culprits.

All great crime stories, Raymond Chandler wrote, are stories of redemption.[21]

3

We mustered at the Comfort Inn in Southport, North Carolina, at 5 a.m. US Fish and Wildlife Service biologists John Hammond and John Ellis were dressed in camouflage Gore-Tex jackets. Patty Matteson, in charge of public relations at the Service, was in her uniform; I had heard that some of the local residents considered the flying duck and leaping fish logo to be crosshairs. The fight between the local residents, armed with chainsaws, and the endangered red-cockaded woodpecker had made national news.

The woman behind the motel desk blurted out, "Are you all working with the woodpeckers?"

Matteson flinched slightly. "Yes," she replied, shoulders almost imperceptibly raised. I looked away, browsing the hotel's kiosk: *Small Changes Can Change Your Life. Bible Suduko.*

"I hope you guys keep it that way. They build too much around here." As we walked into the frigid morning, the woman called out, "Come again and save our world."

The motel's retention pond was stretched tight with ice thin as Saran Wrap. We rolled through Boiling Spring Lakes in the darkness: To the west—Sunny Point, the largest ammunition port in the nation. A "bomb train" occasionally passed through town en route to the military terminal. To the east—the Green Swamp, a swath of wetlands and pine so uninviting it once was refuge to Civil War deserters. No hotel, no restaurants—little more than a gas station, school, and city hall. For Sale signs seemed as common as sapling pines.

We dropped Ellis off at the end of a dirt road. The gray light revealed an ashen opening in the pines: just a few scrawny turkey oaks, their thin, bird-foot-shaped leaves withered and brown. All of the older pines had been cleared. "They've moonscaped these lots," Hammond said.

We started our search for dark holes and white tar beards, the sign of a red-cockaded woodpecker, or RCW. *Picoides borealis* is an unusual bird, excavating its nest in live longleaf pines rather than dead or dying ones—an adaptation to the seasonal fires that once swept through these lands. A dead tree is little more than fatwood, tinder. Excavating the firm heartwood takes time, but there's no rush: unlike snags, living trees are likely to stand for decades, even centuries, and woodpeckers often pass their nest and foraging grounds to their offspring.

"We're being eyeballed," Hammond said as we walked down a side street. Hammond had close-cropped hair, a legacy perhaps of his first thirteen years with RCWs, spent at Camp Lejeune, a Marine Corps base camp about 70 miles to the northeast where he monitored about 60 clusters of them.

"The old hairy eyeball," echoed Ellis, who joined us after he had located the ones he was looking for.

"That's what good neighbors do, of course," said Hammond.

"Except you all are being creepy," Matteson joked, "out looking for birds."

Many people liked having their pines. Fewer appreciated having federal officials in the streets, dressed in camouflage, binoculars round their necks, GPS units at the ready. "Why don't you go look in your own backyard?" someone had snarled at Hammond as he glassed their trees one morning. "I have and there aren't any," he wished he had replied. Another had asked him about a hole in a tree, smooth and circular, just big enough to swallow a child's fist, that turned out to be an active RCW nest. "Leave the nest," Hammond had told the landowner. "You can't cut it down." The cavity was still there the next day, but the crown of the tree had been sawn off. The tree died; the bird moved on. No charges were filed. Hammond laughed at the brazenness as he told me the story.

We spent about half an hour in one lot, ground-truthing the trees that had been marked out on a recent building application. Ellis and Hammond measured distances with a yellow tape, walking along the road and examining each of the mature longleaf pines. The youngest trees blended into the surrounding wiregrass, growing slowly, lying low. In their first few weeks, the young pines had grown taproots about twenty inches long. By the end of their first year, they still looked like grass, but the root could extend eight feet beneath the surface, enough to reach the deep waters beneath the sand and set an anchor for the future tree. Working underground made sense; the pines were vulnerable to fire above the surface. After a couple of years, they would enter the candle stage, a foot-long trunk with a green flame. At about ten, they would resemble miniature palms. By the time they were teenagers, gawky and long-haired, the trees would be growing four or five feet a year. Their foot-long needles looked torn out of Dr. Seuss.

Wind dashed across the canopy. "They really do whisper," murmured Matteson. "You don't feel like you're alone out here." I found myself breathing almost in synch with them. There were still a few that had been permitted to be cleared before the restrictions went in; it was only a matter of time before these would be cut down. But just about every large tree had been marked with red flagging, prohibiting felling. Only the smaller ones, the ones that could be taken out, were ringed with duct tape or slashed out on the property map in Hammond's hands.

Once there were 92 million acres of longleaf pine along the Atlantic and Gulf coasts. With millions of trees measuring a yard or more in diameter, it was "one of the most wonderful forests in the world," said a leading ecologist in the 1930s.[1] The open barrens seemed as vast as the sea; blazes had to be cut in the longleaf to mark even well-traveled paths. "It was really like navigating by means of the stars over the trackless ocean," Basil Hall, a retired British naval officer, remarked while traveling from Savannah to Macon in the late 1820s.[2] Fanny Kemble found the road through Georgia "a deep, wearisome sandy track, stretching wearisomely into the wearisome pine forest."[3] (She was an actress.) After slogging through the morass of the hardwood bottomlands, Union soldiers were said to have found their spirits lifted when they reached the open, park-like setting of the pines. When the war ended, Northern doctors sent their consumptive patients to these woods to escape the toil and turmoil of city life. One doctor wrote, "That the exhalation of the pine is directly healing to the diseased lung is no longer doubted."[4]

The forest stood on flames. Hall described a longleaf fire as "an exceedingly pretty sight. A bright flaming ring, about a foot in height, and three or four hundred yards in diameter, kept spreading itself in all directions, meeting and enclosing trees, burning up shrubs with great avidity, and leaving within it a ground-work as black as pitch, while everything without was a bright green interspersed with a few flowers."[5] Fire in these forests, Lawrence Earley wrote in *Looking for Longleaf,* is like rain in a rainforest: the defining evolutionary force, keeping the trees well-spaced, the understory cropped.[6] The fires could travel slowly, "as gentle as the whisper of a breeze," wrote Edward Komarek, Sr., director of Tall Timbers Research Station, or with tremendous speed.[7] These shin-high fires were wind-driven, commonplace, and often huge. In 1898, a fire blackened three million acres in North Carolina. The newspapers found it hardly worth mentioning.

As I drove through the South, songs on the radio resonated with the forest: The haunting "In the Pines," where the sun never shines and the cold winds blow. The moonlit "Cabin in Caroline," with the blue-eyed girl and the whippoorwill. Hank Williams's long-gone whistle of the old "Log Train." To many nineteenth-century landowners—and to Southern states—the forests amounted to little more than board feet. Land was swapped at costs that now seem astonishingly small. During Reconstruction, Northerners bought up enormous tracts of Southern land; one legislator purchased more than a hundred thousand acres in Louisiana in 1876 for less than a dollar an acre. In 1881, the State of Florida sold off four million acres of forest to a Philadelphia company for 25 cents an acre.[8] Lumbermen often cut and ran: land with standing pines was taxed at a higher rate than cut-over land. And once they started, they had to move quickly to beat the fires fueled by the tinder they left behind.

By the twentieth century, most of the longleaf were gone. "The complete destruction of this forest constitutes one of the major social crimes of American history," a couple of ecologists wrote in the 1930s.[9] Three million acres remain, about 3 percent of the original forest. There are only a few scattered remnants of old growth, the largest at Eglin Air Force Base on the Florida Panhandle, where about ten thousand acres stand, including a tree that is five hundred years old. What once was a rich carpet of longleaf extending from east Texas to Virginia is now just a few ragged tatters.

Some call our epoch the Anthropocene, as human influence is now planetary in scale. Others have dubbed it the Catastrophozoic or the Homogecene. Daniel Pauly, a fisheries biologist at the University of British Columbia, proposes the Myxocene, from *muxa,* the Age of Slime.

If you take the short view, the epoch started around the eighteenth century with the Industrial Revolution. David Ehrenfeld gave the time that humans began to believe they could change the earth a specific date: November 16, 1869, the day the Suez Canal was opened to traffic, shortening the route between India and England by six thousand miles and changing the functional geography of the Old World.[10] A longer view pushes the epoch back at least ten thousand years to the moment humans started changing the land through agriculture and the earth's species through domestication, casting a hominid shadow over the entire interglacial period known as the Holocene.

We are products of diversity: humans emerged in Africa alongside the greatest number of species ever to share the earth at a single time. The earliest records of modern *Homo sapiens* date back about 130,000 years, the time of mitochondrial Eve and perhaps Adam Y, our genetic ancestors, a small group of closely related gatherers and hunters. At least one study indicates that humans (all the *Homo* species) were at such low population levels about a million years ago that they could have been considered endangered, on par with contemporary gorillas and chimps.[11] However, it didn't take us long to make an impact. People followed the coast around the Horn of Africa to the mouth of the Red Sea about a hundred thousand years ago. We settled around the Mediterranean, littering the coast with middens. Some species recovered from overharvesting; many did not. *Tridacna costata,* once the most common giant clam in the Red Sea, all but disappeared. Its formidable foot-long shell, with a zigzag opening resembling a leg trap, did nothing to save it. It didn't help that the giant clam lived along the easy-to-reach reef tops. It is now so rare that researchers didn't describe the species until 2008.[12]

More than fifty thousand years ago, we moved through the Middle East into Asia. Time started to fly: humans settled the coastlines of southeastern Asia and Indonesia, arriving in the Antipodes about forty-five thousand years ago. We took Australia by firestorm, burning down the native forests and hunting ten-foot kangaroos and giant land tortoises. Species with rich and varied diets, accustomed to a life in the trees, now had to survive on shrubs. *Genyornis newtoni,* the tall flightless thunderbird, went extinct soon after people arrived. Mostly, the small survived; all of the species that were bigger than man, or greater than a hundred kilos, disappeared.

Until about thirteen thousand years ago, North America had more large animals than present-day Africa. There were 56 species of herbivores that weighed more than 30 kilograms, or more than a German shepherd. At least half were as big as a moose. Mammoths and mastodons dwarfed any of the species alive on the continent today. There were giant sloths and giant tortoises. Fifteen species of carnivores were at least as big as a coyote: the American lion, saber-toothed tiger, dire wolf, and giant running bear, along with the jaguar, wolf, and mountain lion. There were so many meat eaters, it was tough to find enough food, even on the mild southern end of the Pacific coast. Many of the large cats and canids found in the La Brea Tar Pits had fractured canines and carnassials, the teeth used for shearing flesh. They were eating the carcasses down to the bone. Modern carnivores can be more selective, often keeping their teeth intact into old age.[13]

Our arrival on the continent more than fourteen thousand years ago was followed by the extinction of mammoths, mastodons, giant sloths, and giant beavers. Even the apex predators (the short-faced bears weighing more than a ton) soon disappeared on the heels of their outsize prey. Only 18 of the original 71 mammals bigger than the domesticated dogs that followed us over the land bridge survived. If you're reading this in the eastern United States, your idea of local megafauna is probably down to a single species: white-tailed deer, an edge-loving, carefully managed browser. There may be an occasional black bear, moose in the north, or a recent arrival, the coyote. In the West, your options are slightly wider, at least if you're near a national park: bison, mule and white-tailed deer, elk, pronghorn, grizzly and black bears, mountain lions, and now wolves. Beyond the parks, often only the deer and antelope survive.

Archaeologists still vigorously debate the direct cause of the collapse, but any doubt that humans played a role is becoming less and less reasonable. Driving through the Bighorn Basin of Wyoming, I happened to pass a roadside marker at the site of a steep arroyo, a Clovis killing ground. The Clovis, among the earliest humans in the Americas, would have found mammoths, horses, camels, and bison when they arrived. Did these hunters have the tools to take out many of these Late Pleistocene giants? Piles of bones suggested that they cached mammoth meat during the winter. Were they scavengers or did they work the arroyos? In the 1980s, George Frison, founder of the Anthropology Department at the University of Wyoming, traveled to Zimbabwe with replicas of the Clovis spears, the distinctive, difficult-to-knap point sinewed to a shaft of wood. He and his colleagues tested the weapons on African elephants. (It was the eighties: only dead elephants were used in the experiment.) Thrown from a distance of 65 feet, the six-foot spears deeply penetrated the back and ribcage, breaking only if they hit a rib. Frison concluded that the spears could kill a mammoth if thrust through the lungs. Ancient butchering tools worked efficiently on elephants.[14] So the Clovis could and did kill mammoths and mastodons. But did they kill them off?

Frison thought that hunters couldn't take on an entire group or a full-grown mammoth, so they stalked young careless strays; other archaeologists believe that the Clovis isolated entire family clans.[15] Opportunistic attempts could be costly; angered mammoths were also armed (with eight-foot tusks) and towered over the hunters. Some biologists proposed that perhaps it was global warming combined with changing habitats that took out many of the large animals. A recent study of the fungi growing in mammoth dung suggests that the decline of mammoths occurred fourteen thousand years ago—before the climate shift and before the vegetation changed—just as human hunters were spreading through the continent. It wasn't habitat change that killed the mammoths: rather, the death of the mammoths changed the habitat, then the climate.[16]

Now we're entering the Holocene, creeping close to modern times. Unusual species disappeared, as did the common. Beaver were hunted in Europe for the hat industry until they went extinct in the seventeenth century. Then the Europeans turned their attention to the Western Hemisphere, employing and conquering the Native Americans, in part to gain a regular supply of beaver hides. Near annihilation of the animal followed—and the hydrology of North America changed. Beaver ponds trapped and stored sediments and recycled nutrients; they kept streams warmer in winter and cooler in summer. Without dams, the flashiness of rivers increased: they became more flood-prone, and erosion brought the waters more quickly to the sea. Only now are the rodents and their ecosystems starting to recover in the United States. Passenger pigeons (once perhaps one of the most numerous birds on Earth) were baked in pies, loaded in bulk on schooners, sold at a penny a piece, and used as target practice and hog feed. By 1900, they were extinct in the wild; a decade later we were down to a single bird. Humanity plundered on, unfazed—though some suspect a connection between the pigeons' decline and the rise of Lyme disease.[17]

For many, 1600 is a benchmark year; then the age of extinction began in earnest. The whole idea of mass extinctions, of extinction at all, is relatively new. The Latin verb *exstinguĕre* means "to kill a flame." The earliest use of the word to describe the last male in a family line dates to 1581, the same time (as ornithologists Jerome and Bette Jackson point out) that the dodo was discovered on the island of Mauritius.[18] But naturalists rejected the idea for centuries. Lamarck, who coined the term *biology* in 1802, denied extinction occurred, arguing that organisms resembling strange fossils simply existed elsewhere on the planet.[19] "Such is the economy of nature," Thomas Jefferson wrote, that "no instance can be produced of her having permitted any one race of her animals to become extinct; of her having formed any link in her great work so weak as to be broken."[20]

Yet species were disappearing. In 1765, John and William Bartram had discovered a tree along the Altamaha River in Georgia. It wasn't in flower, so father and son weren't certain about its identity. (John, the preeminent botanist of the age, noted "severall very curious shrubs" in his diary.) When his son returned to the site a few years later, he found the late bloomer in full flower, which he described as being of "the first order for beauty and fragrance." William and a cousin described the plant, giving it the Latin binomial *Franklinia altamaha* (the species epithet from its type location, the genus after Benjamin). He sent two Franklin tree seedlings to France and kept two for his garden. By 1803, the last wild *Franklinia* had been lost to cotton fields, floods, fire, or perhaps fungi (no one knows for sure). It was the first known American plant species extirpated by humans. The seeds Bartram collected for his Philadelphia garden—the ones sent to France disappeared—are the only reason the plant, extinct in the wild, has not been lost entirely.[21]

The French naturalist Georges Cuvier may have been the first to demonstrate that species had gone extinct in the past, coming to this conclusion while examining fossils in the gypsum mines of Paris. In *Revolutions on the Surface of the Globe,* published in 1812, he wrote, "Life on this earth has often been disturbed by dreadful events. Innumerable living creatures have been victims of these catastrophes."[22] His meticulous studies of living and extinct elephant species soon eliminated any reasonable doubts. Darwin, aware of Cuvier's work, argued that complete extinction was a "slower process" than the production of a new species. Extinction occurred in ripples, he believed, rather than cataclysms.

If only.

In the 1990s, Robert May, then president of the Royal Society of London, and colleagues compiled a comprehensive list of known extinctions since 1600: 485 animals and 584 plants.[23] Not much more than a thousand species. That's much higher than the background rate in the fossil record, estimated at about one species in a million per year, but it's not exactly through the roof. The trouble is, except for the most conspicuous species—trees, flowering plants, and large vertebrates such as mammals and birds—we don't really know what we're losing.

When May put the list together, there were only two known amphibian extinctions. Now at least 122 species have gone missing (though the Red List's EX may not yet be beside all of their names).[24] More than a thousand golden toads were recorded in the breeding ponds of the Monte Verde Cloud Forest Reserve in Costa Rica in 1987. The park was well protected and well studied, the frogs as cute as amphibians get, yet two years later researchers couldn't find a single individual: the toad hasn't been seen or heard from since. How did this happen? No one knows for sure. An invasive chytrid fungus, a thinning ozone layer, and local shifts in climate have all been floated as possible causes. But such a stunning decline in just 48 months caused panic among some herpetologists. What at first appeared to be a temporary dip in amphibian populations—there was little long-term data for many tropical species—began to look like a final descent: 20 species of Costa Rican frogs and a dozen Australian species took a tailspin at the same time. At least 30 of the 110 species of

Atelopus harlequin frogs have gone extinct, 29 of them in the past 20 years. The total loss may be twice as high.[25] In a race against time, zookeepers have started collecting frogs throughout the tropics. Whether there will be anywhere safe to return them to remains an open question. There is the danger that we have brought them inside, like cut flowers, simply to watch them die.

The consensus among biologists is that the extinction rate for all species is now fifty to five hundred times higher than the long-term average.[26] Others put it an order of magnitude higher. The large-bodied animals are especially at risk, prompting some to ask: has the Cenozoic, the Age of Mammals, come to a close? A comprehensive study, led by Jan Schipper at the IUCN, with 129 coauthors, revealed that more than a quarter of the 5,487 mammal species that have been described are at risk of extinction. Four out of five species of apes and monkeys in southern Asia are in immediate danger, and more than a third of marine mammals.[27] Some researchers estimate that the number of threatened birds and mammals will rise by 7 percent by 2020 and by 14 percent before 2050.[28] Others fear that climate change and habitat loss may increase the current pace tenfold.[29]

In a sense, this is an extinction echo, fueled by fossils from the Permian or the end of the Cretaceous. Our constant motion has created a new Pangaea, a supercontinent connected by petrochemicals. Of course, we aren't running the extinction crisis on petroleum alone: we are also burning our forests alive. In the 1990s, about six million hectares of humid tropical forest were cleared each year, and a further two million degraded.[30] At current rates, a sixth of the world's existing tropical forests will disappear by 2050, and almost a third by the end of the century. Loss of forest leads to drought: less water evaporates from the cleared land and fewer raindrops fall. The remaining soil bakes in the open sun, forming ground as hard as concrete, which can lead to flooding when the rains finally do arrive. Ex-forest is flood-prone and drought-prone, the vegetation useless as microorganisms disappear and growth slows.[31] New ecosystems will take their place, giving rise to unprecedented combinations of native species and exotics, new climates, and a reduction in ecosystem services such as primary productivity and the provision of fresh water.

According to the Red List, one in three amphibians, one in four mammals, and one in five bird and plant species are at risk of extinction. More than half of the snake species examined in one study have experienced sharp declines in the past twenty years.[32] Another study, using "electronic lizards" in Mexico (constructed from PVC pipe, automobile primer, and thermal sensors), has shown that if rapid action isn't taken against climate change, it could soon be too hot for many lizards to forage during the day; one in five lizard species might go extinct by 2080.[33]

Species-level statistics are extremely troubling, but they fail to tell the whole story. Each year, about one out of every hundred animal and plant populations goes extinct. One out of a hundred! Andrew Balmford at Cambridge University and colleagues have estimated that we could see the collapse of 15 to 35 percent of all wild populations in a single human generation.[34] The situation is bad in the forests, worse in the oceans, and perhaps worst of all in freshwater. We could lose all the skinks on a ridge in Florida, a cave of bats in Europe, a reef of deep-sea coral, a troop of gorillas in the Congo, a bed of mussels, and a forest in Africa—clear-cut, with all the mahogany gone.

As populations disappear, a species loses genetic diversity and, in some cases, cultural memory. Entire cultures of grizzlies, with their own behaviors and feeding traditions, have disappeared from much of North America. Thanks to commercial whaling, northern right whales no longer calve in the Bay of Biscay or feed off the British Isles. Fin whales were extirpated from the feeding and calving grounds along the Straits of Gibraltar in the 1920s. Blue whales were hunted to extinction off the coast of Japan by the 1940s. Humpbacks have been extinct off South Georgia in the South Atlantic since 1915. All group memory of the habitat, the intricate knowledge of local waters and unique whale dialects, is gone.

In the three years since I started writing this book, several species have disappeared entirely, or researchers have given up the search for them. Two critically endangered frog species were declared extinct in Costa Rica, even though, well protected in a national park, both had been abundant throughout their ranges as recently as the 1980s. Not a single breeding male of either species has been heard in this century. The white lemuroid possum was declared extinct in 2008 after extensive searches in its native mountain forests in Queensland turned up no sightings. (Three individuals were later discovered in March 2009.) The Liverpool pigeon of the Pacific Islands, unseen for more than a century, was finally declared lost. Field surveys in French Polynesia confirmed the fears of invertebrate biologists: most species of land snails in the genus *Partula* were gone. In 2009, the Red List raised to 51 the number of these endemic tree snails now extinct, victims of an invasive predatory snail brought in from Florida to control a crop pest—another introduced species, the giant African snail. Two of the *Partula* species were used to make shell leis, a tradition now lost. Forty-eight field surveys and 2,500 kilometers of transects failed to turn up a single western black rhino in northern Cameroon; the entire subspecies is probably extinct. The Alaotra grebe, last seen on a Madagascan lake in 1985, was declared extinct in 2010. All that remains is a single blurry photo and a few museum specimens.

In junior high I was taught that all narratives boil down to four basic struggles: man versus nature, man versus society, man versus man, and man versus himself. In nature, the struggles for existence have often been distilled to the classic Darwinian conflict of organism versus organism, leading to competition (which favors speed, size, thorns, or poisonous compounds to defend against hungry browsers). It's Roadrunner versus Coyote, Spy vs. Spy, an escalating arms race that biologists have likened to Lewis Carroll's Red Queen, desperately running just to stay in place. Once the predator has been hunted off the island, the armor looks pretty odd. The pronghorn's speed was essential to escape North America's prehistoric cheetahs. The thorns of the honey locust tree were effective against mammoths. The thorns and dash are useless now, but the defense lingers on.

When organisms face the environment instead of one another, it can lead to an entirely different strategy: mutualism. Cooperation, or "mutual aid," had been largely overlooked until Kropotkin, the famed anarchist, geographer, and naturalist, compiled writings about his Siberian travels in 1902. Rather than simple Darwinian struggles, Kropotkin saw far more evidence of the "mutual dependency of carnivores, ruminants, and rodents."[35] Species work together to overcome the harshness of the steppes, the Arctic, or deep-sea vents, where mouthless worms house bacteria that convert methane to energy. Why struggle for food when an internal colony of asexually reproductive microbes can provide all the energy you need from the methane-rich waters around you?

So here's the deal: for much of our evolutionary history, it was man versus man or man versus tiger. Now, we're up against a changing world—and in a sense we're up against ourselves. We're going to need every organism we can muster in the struggle.

"Junco," John Hammond responded to a nearby trill, not looking up from his Garmin as he GPSed a starter hole in a tree. It was a work in progress, a bright oval a few inches deep, fifteen feet off the ground, with a white tar beard. Red-cockaded woodpeckers work on several cavities at a time, chiseling their way through living, resinous tissue, excavating upward until they reach heartwood. Then they burrow down, creating a spacious capsule. Usually it takes two or three years to complete a nest, though sometimes it can take five or more. When the sap stops oozing in the cavity, the birds move in. Trees are typically passed from father to son, so a young male might stick around, helping to excavate cavities and rear his siblings: if he's lucky, he'll inherit the property. Most daughters set off alone soon after they fledge—but they don't fly solo for long. As one biologist told me, "A solitary female does not happen."

"Carolina wren," called Hammond. Even I recognized the *Tea-kettle, teakettle, teakettle.* The sun had finally begun to burn through the clouds.

There were long gashes at the base of the tree, a catface. Metal emerged from the bark like shark teeth, a legacy of the nineteenth-century tar heels. Longleaf, the most resinous of Southern pines, was a leading source of turpentine, a household cleaner and cure-all: the Romans used turpentine to treat depression; naval surgeons used it to dress wounds and amputations; physicians fought tapeworm and bronchitis with it. The American navy would have sunk without it: hulls were sealed with longleaf pitch to protect against shipworm. Leaks were caulked with oakum and pitch. Ropes on sailing ships were coated to preserve them from salt water. Even canvas cloths were tarred: tarpaulins. Punishment at sea was meted out with pine tar and feathers.[36]

Turpentine and rosin came from living trees; tar from dead ones. To gather turpentine, the trees were boxed, a pocket chopped deep in the base. The resin dripped from long shallow Vs above the box and was collected every week or two, the gum ladled into barrels and taken by oxcart to the nearest still. Only first-year trees yielded the virgin dip, the most valuable, clearest rosin. The longer a turpentiner worked a stretch of forest, the less he made. After the fatwood was burned and the tar gathered, the charcoal that was left was sold to blacksmiths.

The resin is so sticky that even red-cockaded woodpeckers can face death by flypaper if they're not careful. The slight risk is worth it: this candling works. In the 1980s, Craig Rudolph and colleagues at the US Department of Agriculture in Texas ran experiments on rat snakes, an abundant and efficient predator of birds in pine forests.[37] The dark snakes were released a couple of feet above the base of a long-leaf, at the lower end of a sixteen-foot climbing course: some trees were covered in resin, and some had the scaled bark removed, while the unaltered ones acted as a control. Almost all of the snakes climbed the intact and scaled trees, though some refused to participate, usually by falling or slithering toward the ground. (These were gently prodded with telescoping fishing rods.) The lack of bark slowed the snakes down by removing crevices that helped them gain purchase on the tree. The resin stuck to the ventral scales, the scales stuck to each other—and a tarred snake can't climb. Only three of eighteen snakes made it to the tarred cavity nests. Candling, which can run twelve feet or more down a tree, is the woodpeckers' best defense. It may also serve the same purpose for the birds as it does for birders: a white beard in the middle of the forest can be seen from a distance. If it's active, it means only one thing: red-cockaded woodpeckers.

For most of its evolutionary history, the red-cockaded wood-pecker's nest probably had little effect on the health of the longleaf. The birds survived on grubs and every stage of arboreal ants—eggs, larvae, pupae, adults. The trees lived long after the birds abandoned the nest; often new tenants would move in, such as owls, flying squirrels, or other species of woodpeckers. The trees' ability to endure the constant loss of sap from the resin wells and the bark-scaling may even explain why many catface trees are still standing more than a century after the scrapes from the tar and turpentine industry healed over. Now a woodpecker's nest can keep a tree alive—and prevent it from being sawed to the ground even in the midst of a booming city.

As we passed a mature pine, cut up and dumped on the lot, Hammond asked, only half in jest, "Y'all want to see if there are any cavities on that tree?" If a tree falls in Boiling Spring Lakes, it's a potential crime scene. "At least two active trees have been cut," he said. "There may have been more, though they were probably inactive. It's hard to prove, especially when the logs are gone."

Two red-cockaded woodpeckers rushed into the lot. The male, his black wings dabbed with white spots, landed on a turkey oak to forage on the thin crooked branches. The female worked a snag, tapping her bill on the bark, peering underneath. Hammond whispered, "We believe it's a breeding pair." On cue, the male swooped down and mounted the female.

"You can have a cavity tree right next to a house," Hammond said, "but they have to have somewhere to eat. You don't want too much cement."

A man with a resinous white beard walked up the dusty road. "Seen any good woodpeckers?" he asked. Matteson hesitated, then told him about our morning.

"I tried to buy this land from Fred Hicks," he said, "who bought it in '64, but I can't find him. I'd love to keep it just as it is."

We drove to another of the many sites with permits pending. Shareen Hummel, a real estate developer, was pacing when we pulled up.

Ellis looked over the lot. It had a couple of longleaf pines, but no sign of birds. He thought he could draft a letter that would allow Hummel to build without a permit.

"I don't want to get put in jail," she said.

"It's a dialogue," Ellis told her. "If your plan changes, come talk to us again."

"I just don't want to purchase these and then be stuck."

"Once you get that letter, you can do what you want."

We followed Hummel back to her five-acre residence on the outskirts of town, where she was hoping to build a few homes. I noticed a few catface trees.

"Lots with trees definitely have more appeal," Hummel admitted. "I think they add value. But I can't buy lots I can't build on. This place started to boom a few years ago. Now nothing's happening."

It wasn't until after Hummel had bought the place, she told us, that her neighbors had asked, "Oh, by the way, have you heard about the woodpeckers?"

The next day, she found one across the street.

"There's going to be take in Boiling Spring Lake," Matteson said. *Take* in the parlance of the Endangered Species Act could include the direct and indirect killing of protected woodpeckers. It is illegal to harm or kill an endangered species, or to harass, pursue, hunt, shoot, wound, trap, or capture one. You also can't modify a listed species' habitat, on public or private lands, if it will indirectly lead to its injury or death, *incidental take*. Unless you have a permit. "There are places that grow woodpeckers," Matteson said. The city had contracted a group to draft a Habitat Conservation Plan, a process that would allow some areas to be developed as long as the total number of birds stayed the same.

As we walked her lot, Hummel hid nothing. "I'm sorry I can't find the holes over here, but there were some." The snags looked like gnawed cobs. Yellow-bellied sapsuckers had drilled vertical dotted lines in the bark, exposing the red wood beneath. Sphagnum moss covered the ground. "These pinecones drive me crazy," she said. She had paid some kids to collect them. The next day there were more all over the place. "I know there was a hole around here somewhere."

Hammond eventually found a shallow one on an older pine.

"You should probably talk to Jenna pretty quick," he told Hummel. Jenna Begier was the state's Safe Harbor biologist, in charge of habitat permits for the endangered woodpeckers. If Hummel registered now, her property would be registered as a Zero, no nesting birds, and she would face fewer restrictions. "If Boiling Spring Lakes goes to full build-out," Hammond continued, "the Safe Harbor Program could make these trees an investment opportunity. You could sell your woodpeckers to other developers."

I was taken aback by Hammond's comment, suggesting that Hummel get into the program *before* the birds moved onto her property. The woodpeckers might not live there yet, I thought, but they were working on it. Why rush landowners into a program that would give them the option to cut down trees, even if woodpeckers showed up? Hammond was supposedly working to protect the birds. Was this what the Fish and Wildlife Service had come to?

As we drove away, we passed a small fort. Two boys rose up from behind the cardboard and started shooting us with their toy guns.

"Isn't that cute?" Matteson said.

In the 1990s, two California scientists, Michael Soulé and Michael Gilpin, plotted the environmental and biological factors that can push rare species into what they described as "extinction vortices." Each vortex—disturbance, fragmentation, loss of genetic variation, or an increase in genetic drift—can be enough to drive a species to extinction.

The two most dramatic factors, and the easiest to observe, are external. An asteroid slamming into the Yucatán, a superstorm, an abrupt shift in climate. But disturbance can happen on a much smaller scale, too: all it took to wipe out the Sampson's pearly mussel were a few dams on the Warbash and Ohio Rivers, which silted up its gravel and sandbar habitats.

Small populations are subject to sheer chance, the rolling of the dice, or what conservation biologists call *stochasticity*. Working on tiny flour beetles in their UC Davis lab, Brett Melbourne and Allan Hastings have shown that when a population dwindles to just a few individuals, demographic stochasticity—the ratio of males to females, for example, or the chance that a few of those that remain are not very good breeders—can outweigh environmental factors. For most species, sex is randomly determined, and sex ratios fluctuate in small populations. The extinction risk goes up if most of the population is female or, worse, male. Differences in the behavior and reproductive output of individuals can also increase the chance of extinction. The Seychelles magpie robin, down to just seven breeding pairs in 1988, forms social groups of one dominant breeding pair and several subordinates that don't mate until the top birds die. Some males may sire most of the offspring while others father only a few or none at all. This reproductive skew reduces the number of breeders, and greatly expands the territory needed to raise just a few fledglings. On dense islands, this behavior might have prevented all the birds from nesting at the same time and fighting over resources. But it has slowed the restoration efforts of this critically endangered bird, increasing the extinction risk.[38]

Finally, there are internal failures: the organism versus itself. We're trained to praise an ability to overcome all odds, but as the number of individuals declines, variants in DNA can become a time bomb. In a large population, slightly deleterious alleles have little impact: they're so rare, and so rarely inherited from both parents (which would unmask their effects) that they never become common enough to impact the species. But when numbers start to dwindle, a once-uncommon allele can become fixed in the population in a process known as genetic drift, leading to loss of reproductive fitness and perhaps extinction.

When does the final decline begin? Depending on the organism's life history and ecology, it can start when only a few dozen or several hundred individuals remain. For one population of red-cockaded woodpeckers, the vortex to extinction kicked in at 19, for wild dogs it was 26, and for the Snake River coho, 404.[39] At this point, population increases did not occur or were very short-lived. On a graph, the slope to extinction looks like a water slide; as populations decline, the extinction rate accelerates.

The impact of the vortex doesn't stop at a single species. When an ecosystem reaches an unstable state—perhaps because new grazers have been brought in, nutrients have been added, or overfishing has removed a trophic level—even a minor disturbance can push that system to a new state, one that is stable, if unfamiliar. Theorists have called this transition a "catastrophic fold," when a change can be rapid and all but irreversible.[40] The fold describes a region of instability—like bending a piece of paper to provide a shortcut between two outer edges or states. I think of Madeleine L'Engle's *A Wrinkle in Time*, where two ends of a string are brought together, allowing an ant to crawl across. Such changes can be rapid (a meteor) or slow (an invasion of a new group of organisms over a land bridge). Either way, the results are dramatic. An algal bloom erupts in a clear lake, depleting the oxygen and turning the water eutrophic, resulting in fish kills. An overgrazed grassland becomes desert scrub.

Extinction will be our longest-lasting legacy. But what happens after the crisis? You might think that a clean slate following a mass extinction would provide the best opportunity for new species to arise—that there would be an explosion of evolution in the absence of diversity. James Kirchner at Berkeley and Anne Weil of Duke University examined marine invertebrates and several plant and animal families, looking for patterns of new speciation in the fossil record. They found that the origin of new organisms increases along with diversity. Old species contain the raw material for new ones, each taxon representing a new evolutionary pathway. Extinction doesn't fuel evolution; it chips away at its raw materials—the rich fabric of DNA and the unique ecosystems and niches. A depauperate earth—the one we're creating—leaves little opportunity for speciation.[41] Kirchner and Weil observed a lag of several million years after major extinctions before the engines of evolution revved up and began to repopulate the planet with new species. The precipitous drop we're causing in the number of species, genera, and families is very likely to outlast our presence here. Stephen Meyer, director of the MIT Project on Environmental Politics and Policy, has already written the epitaph: "The broad path for biological evolution is now set for the next several million years. And in this sense the extinction crisis—the race to save the composition, structure, and organization of biodiversity as it exists today—is over, and we have lost."[42]

Dressed in a sweatshirt and running shoes, Marty Kesmodel, mayor of Boiling Spring Lakes, escorted me to the office of David Lewis, city manager. Woodpecker-foraging maps lined the cinder-block corridors and offices of the city hall. Green and purple circles were bunched over the city like grapes, 25 bubbles that some developers feared would burst their plans. Each bubble represented a foraging partition a kilometer in diameter. The center of the cluster, where the birds nested, was a wine-stained bruise. There were streets named after constellations, trees, crests (Fieldcrest, Pinecrest, Crestview), and a surprising number of birds: a Pheasant, a Dove, a Quail, a Bobolink, a Bluebird, a Seagull, and even an Audubon Road. (But I couldn't find a Woodpecker Drive.) The mayor and the manager eyed the foraging map as if it were a battle plan.

Kesmodel gave me the history. First the developer Reeves Telecom, which had owned several TV stations in the Southeast, decided to subdivide its holdings in 1962, using their stations in West Virginia and the Carolinas as a marketing tool. "A dollar down and a dollar a week would get you a lot on the coast," Kesmodel said. People from nearby Fort Bragg bought property to build on after they retired from the army. In the 1980s, people in the Northeast discovered the low tax rates, and the population changed from retirees to young working couples. Then at the beginning of the twenty-first century, real estate prices spiked and the rush was on. But some of the properties didn't percolate; they were too sandy for septic systems. In a town without sewers, they remained unbuildable. The Nature Conservancy began purchasing the most ecologically valuable lands.

Kesmodel and Lewis tried to downplay the town's response to the red-cockaded woodpecker. But the numbers were clear: during a six-month period in 2005, typical for the time, 20 cutting permits were issued. When the Fish and Wildlife Service put the town on notice that it would soon need written permission to clear lots, 368 cutting permits were issued. Eighty-nine acres were cleared in nine months. Kesmodel asked me if the city looked as bad as the out-of-town newspapers made out. It still looked pretty forested to me, but I hadn't seen Boiling Spring Lakes in the 1980s, before major development began, and I had no idea what a pristine longleaf pine ecosystem looked like. No one else does, either.

Since regulations were put in place, 171 permits had been issued. According to the mayor, only one application had been denied. It was hard to tease apart the national trends—the housing market had bubbled up, then burst during the years since the first foraging maps had been drawn—from the effects of the act, but the spike was undeniable. More permits were issued during the three seasons when the town was put on notice than had been cleared before or since.

Kesmodel couldn't provide any figures on the cost of the Endangered Species Act to his city. "There's an intangible cost, I guess, in the government's taking control of private property, taking the rights of those individual property owners away from them. There's been a strong feeling here of 'Why us?' Why did they decide that we're going to go down to Boiling Spring Lakes, North Carolina, and make *them* comply and not the other communities in the county? That's a cost in credibility, as well as a cost in economics. But with The Nature Conservancy, the benefit is that it seals this area as an eco-area that people can have—the preserve is going to be there indefinitely. And that's a big, big plus, and a lot of people feel that way."

"I can go out there and measure trees," Lea Anne Werder, a real estate agent who had moved from Buffalo a few years earlier, told me later that day. "But you have to go through a lot of additional work to get clearance from Fish and Wildlife.

There were nineteen clusters of red-cockadeds and dozens of requests for new homes. Each woodpecker group needed at least 75 acres of foraging habitat. When I visited, the city was developing a Habitat Conservation Plan that would balance the interests of the developers with the needs of the birds. Most human families were content with quarter-acre lots. But, of course, most didn't forage on their lands. We humans, or at least we Americans, use about 22 acres each per year—for natural resources such as food, timber, and built-up land. That's our ecological footprint.[43]

"I think that the people that lived here for a long time were devastated by the clearcutting," Werder said of the response to the Service's notice. "Aren't they beautiful birds?"

A char-free longleaf pine is a tree in danger. Too much duff (the tree needles and other debris that gathers on the ground) can stop the regrowth of young trees and wiregrass, turning a fire-resilient system into a tinderbox, ripe for a catastrophic fire. After fires devastated the Myrtle Beach area of South Carolina, longleaf pine forests emerged relatively undamaged.[44] By lowering the amount of duff, prescribed burns can protect cities such as Boiling Spring Lakes from wildfires. The Nature Conservancy would need to schedule them to keep its forest healthy. When I visited land it had burned three weeks earlier at nearby Green Swamp, I could still smell the fire, the char slithering a few feet up the trees like reluctant rat snakes.

Already there were moguls of verdant wiregrass growing up under the well-spaced trees.

A green canopy can elevate the beauty of a city, raising its value. The trees also reduce air pollution and help manage storm water. The birds themselves can provide pest control. When it's not eating the southern pine beetle (a major pest and favorite food), a single red-cockaded woodpecker can remove up to eight thousand damaging earworms per acre of corn.[45] Its excavations increase diversity. American kestrels, eastern screech owls, southern flying squirrels, bluebirds, bats, and lizards, even rat snakes, will move in once the resin stops flowing. Their long-leaf pine ecosystem is also the home of endangered rough-leafed loosestrife and twelve other species of concern, including the Venus flytrap, yellow fringeless orchid, Carolina gopher frog, and southern hognose snake. The woodpeckers can even attract tourists eager to see an endangered bird. It was clear the city wanted to have it both ways: at the edge of town, a sign said City Limits, Bird Sanctuary.

With a *pish* and a whoop and a bang on the bole, Ryan Garrison clawed at the base of a fifty-foot flattop pine, where a male bird had been spotted entering a nesting cavity on a wooded lot. He banged his field notebook against the tree. When the bird flew straight into the mesh of an extendable net he had just placed over the hole, Garrison reached in and gently grabbed it.

"What are y'all doing?" a woman yelled. She was standing on her deck, awash in a pink housecoat. Hammond walked over. "You're not taking my woodpecker, are you?" He told her we were just banding it. We would place it back in a few minutes.

"Good," she said, "because I love having it around."

"People with houses love woodpeckers," Hammond said when he got back. "To owners of a lot, woodpeckers are Satan."

Garrison blew on the feathers of the bird's head, revealing a bright red cockade. He drew some blood, and spat some tobacco. The bird weighed 49 grams. He passed it to Hammond, who put two light green bands separated by a black-and-white one, the colors of cluster 14, on the right leg. On the left went a dark blue one and an aluminum US Geological Service tag. Wherever this bird flew, observers could now identify it with this tree.

"Hey, it could be worse," Hammond said of their efforts to protect the woodpecker. "It could be a butterfly." Butterflies were easy, I said. I would soon go see a couple of clam species that the governor of Georgia had accused of endangering the lives of his state's children.

Matteson laughed. "Woodpeckers are pretty, but mussels?"

And so it goes.

Once it was banded, I was given the honor of letting the woodpecker go. Shaken up from the nest rattling, the bird-catcher's net, and the several minutes under the pliers, it was still in my palm, light as a few thousand feathers. I placed it gently on the pine. For a moment, it eyed me across the taxonomic divide.

The bird embodied the hope, the surprising boldness of the Endangered Species Act. Without it, this male and many of the red-cockaded woodpeckers in Boiling Spring Lakes (and throughout the South) would have been lost. It may be underfunded and at times mismanaged, but the Act is an unprecedented attempt to delegate human-caused extinction to the chapters of history we would rather not revisit: the Slave Trade, the Indian Removal Policy, the subjection of women, child labor, segregation. The Endangered Species Act is a zero-tolerance law: no new extinctions. It keeps eyes on the ground with legal backing—the gun may be in the holster most of the time, but it's available if necessary to keep species from disappearing. I discovered in my travels that a law protecting all animals and plants, all of nature, might be as revolutionary—and as American—as the Declaration of Independence.

As I opened my hand, the bird hesitated, talons tight on the rough bark. The moment passed. Through the trees, out of sight, he flew.

4

By the end of the 1960s, the geographic and taxonomic limitations of the government's earliest lists of species in trouble were becoming obvious. No invertebrates, no pelagic species, such as deep-sea fishes or cetaceans, were included, and no animals outside the borders of the United States. The international fur trade killed cheetahs, jaguars, and leopards for spotted coats imported into this country. The great whales were almost gone. Whaling nations were even busy downsizing their harpoons: the explosive grenades developed for the enormous fins and blues could render unusable the relatively diminutive minke whale, the last species that was still common enough to hunt.

In July 1970, the Fish and Wildlife Service, in the Department of Interior, proposed listing eight of these now rare whales: several were clearly endangered, the rights and humpbacks, and three were under threat from continued commercial harvesting: the fin, sei, and sperm. But the Defense and Commerce Departments didn't like the idea. Sperm-whale oil was still used in submarines, one of the few truly profitable enterprises left to whalers; the listing would close that business down. The Pentagon knew it couldn't win the argument on economic terms, so it claimed that sperm whales weren't in any *immediate* risk of extinction, even if their populations had been greatly reduced by overhunting.

The director of the Fish and Wildlife Service, John Gottschalk, acknowledged that, technically speaking, the sperm, fin, and sei whales did not belong on the list. But his boss, Interior Secretary

Walter J. Hickel, thought that Gottschalk's reasoning didn't make a lot of sense. The whales were rare and the commercial hunt clearly at fault. He decided to list them despite the Pentagon's objections, submitting the final rule to the *Federal Register* two days before Thanksgiving. His timing wasn't great: a letter that he had written to Richard Nixon protesting the president's Vietnam policy and the shooting of college students at Kent State had just been leaked to the press. The letter made headlines across the country, and Hickel was fired. Smelling blood, the Pentagon and Commerce Departments pressured Buff Bohlen, the second in command at the Department of the Interior, to withdraw the proposal. But it was a holiday weekend, and Bohlen was a passionate environmentalist. "I somehow couldn't get around to it," he told journalists Charles Mann and Mark Plummer.[1] The whales stayed on the list.

It had become clear to Bohlen and his colleagues that the laws passed in the 1960s failed to meet the true needs of endangered wildlife. Congress had directed the Interior to make lists, acquire habitat, and impose some restrictions on wildlife markets, but conservation efforts were required only when "practicable." Legal protection of threatened species extended to the boundaries of national wildlife refuges, but no farther.[2] Citizens could kill any of the species on the original list (unless other laws protected them). They could destroy habitat without risking prosecution. Federal agencies didn't have to consider the list if they wanted to build a dam or highway, even if it caused a local or worldwide extinction.

The Fish and Wildlife Service was doing what it could to protect several species, such as the whooping crane and California condor, but without a congressional mandate, it had to ask for money each year. In 1969, Congress had appropriated only a meager $1.3 million to acquire endangered-species habitats. Even so, the Endangered Species Preservation Act was revised that year to make it a crime to bring species protected outside of the United States into the country and to export native ones. There was a call for an international treaty on the trade of wildlife and wildlife products.

According to John Ehrlichman, the presidential assistant for domestic affairs at the time, Nixon thought that environmental issues had no "political magic. They had the potential for costing a lot of people jobs, and that was bad, but they also were something that a small number of people cared passionately about, so that it was an issue that wasn't going to go away."[3] By 1970, the number—and the passion—appeared to be growing: Wisconsin senator Gaylord Nelson organized the first nationwide environmental protest, a "teach-in" modeled on Vietnam demonstrations. Twenty million Americans participated in that first Earth Day. Congress and the Nixon administration got the message, soon passing the Clean Air Act and creating the Environmental Protection Agency.

By 1972, Nixon was calling for the adoption of a new robust endangered species law "to provide the kind of management tools to act early enough to save a vanishing species."[4] He announced a ban on poisons used to kill grizzly bears, gray wolves, and other rare predators on public lands and submitted a new bill to Congress, drafted by Buff Bohlen and colleagues. (Privately, Nixon told Ehrlichman to keep an eye on the law.) The 92nd Congress never managed to pass the endangered species bill, but legislation that would prohibit the killing of whales and other marine mammals in US waters and stop the import of their products into the country did come up for a vote.

To get the Pentagon to sign on, Lee Talbot, who had done the early surveys for the IUCN in the 1960s, met with representatives of the Joint Chiefs of Staff and the president in a guarded vault deep under the east side of the White House. The military was still concerned about the supply of sperm whale oil, so Talbot carried an affidavit from DuPont stating that it could produce artificial whale oil in a couple of months, though its price was somewhat higher. The White House decided in Talbot's favor, and synthetic oil was produced in about six weeks. The Marine Mammal Protection Act was passed in October 1972.

Talbot, as senior scientist in the new Council on Environmental Quality, had been working with Bohlen on the legislation that would become the Endangered Species Act. He and a colleague on Capitol Hill, Frank M. Potter, Jr., were determined to plug any holes in the early draft before it was signed by Nixon, the struggle over whales having shown both men that too much flexibility was a bad idea. They eliminated the word *practicable* from the legislation, which they believed provided too much wiggle room, and revised the preamble to emphasize ecosystems as well as the species themselves. Neither of them advertised these changes—changes that would alter the course of the Act—and no one in Congress or the Nixon administration appeared to notice.[5]

Michigan Democrat John Dingell approved the revision and introduced the bill in the first session of the 93rd Congress on January 3, 1973. The House passed it 390-12. Later that month Mark Hatfield, a Republican from Oregon, submitted almost identical legislation to the Senate, where it passed 92–0. While the final version was being hammered out, the Convention on International Trade in Endangered Species of Fauna and Flora (CITES) was being drafted in Washington, DC. At the meeting, an announcement by Assistant Secretary of the Interior Nathaniel Reed that federal agents had broken an international ring charged with smuggling more than a hundred thousand spotted cat pelts into New York helped get the international agreement passed, establishing a system of permits and prohibitions on the trade of species threatened with extinction. The United States was the first to ratify it.

When the Act came up for a final vote, there were only four nays in the House and no dissent in the Senate. Supporters included Republicans Bob Dole of Kansas, Jesse Helms of North Carolina, Ted Stevens of Alaska, and Howard Baker of Tennessee. Nixon signed the bill on December 28, 1973, noting the "legislation provides the federal government with the needed authority to protect an irreplaceable part of our natural heritage—threatened wildlife.... Nothing is more priceless and more worthy of preservation than the rich array of animal life with which our country has been blessed. It is a many-faceted treasure, of value to scholars, scientists, and nature lovers alike, and it forms a vital part of the heritage we all share as Americans."[6]

The Act was visionary and comprehensive, a feat just about unimaginable 40 years on. It was prohibitive, with enormous reach: there was to be no "take" of endangered species—not through direct hunting, removal, or the destruction of a species' habitat—a rule that could prohibit cutting trees, clearing land, or diverting a river or stream. Recovery plans under Section 4 of the Act were to be based on conservation science, covering a species' biology, its past and present distribution, and reasons for listing. Later changes would insist that such plans include prioritized measures necessary to abate identified threats, population targets for delisting, and the resources needed to achieve recovery. Sections 5 and 6 authorized land acquisition, financial assistance, and cooperation with the states. Section 7, perhaps the most powerful, required the suspension of all federal actions jeopardizing an endangered species or its critical habitat. This early mention of habitat "critical" to the continued existence of a species has since been expanded to include areas that are important to a species' well-being, even if they are currently unoccupied. Unlike the listing process, such habitat decisions required that economic considerations be taken into account. The Fish and Wildlife Service, instructed to review all the vertebrates that had been listed, retained 131. The eventual delisting of them all just might have seemed an almost achievable goal.

The Act had set out to define the levels of extinction. An endangered species was "in danger of extinction throughout all or a significant portion of its range," a threatened species "likely to become endangered within the foreseeable future." These definitions were models of imprecision, according to Michael Bean, a longtime lawyer at the Environmental Defense Fund and now a counselor in the Fish and Wildlife Service.[7] Congress intended that endangered species would enjoy more legal protections than threatened ones, with the endangered being protected by the full array of prohibited actions. The Service could pick and choose among the rules for those merely threatened, but in practice both groups have been treated about the same. What exactly was meant by the "foreseeable future"? Was it ticked off in years or generations, which vary from species to species? Such vague wording would drive the courts crazy.

But the Act helped push wilderness preservation away from a focus on aesthetics and recreation to one based on wildlife and the science of ecology. It also gave the sometimes-rambling National Wildlife Refuge system direction and cohesion. Since 1973, more than 60 refuges, totaling about 375,000 acres, have been created to conserve listed species.[8] The Oklahoma Bat Caves protect endangered Ozark big-eared and gray bats. The Sweetwater Marsh conserves 316 acres of wetlands in San Diego for the light-footed clapper rail. The 106-acre Moapa Valley NWR in Nevada was established to protect the streams and springs essential to the Moapa dace, a three-inch fish endemic to warm water springs. One hundred and thirty-nine acres of the Ellicot Slough were protected for the Santa Cruz long-toed salamander. In Missouri, the Ozark Cavefish NWR protects 40 acres for the eponymous blind fish. Even small creatures need decent-sized parcels of land.

The Endangered Species Act followed several iconic pieces of legislation in the United States, but the tradition of nature protection really began in the nineteenth century with the creation of national parks. Natural history museums, which had become popular by then, proved enormously influential in developing the idea of biological diversity, not only across kingdoms and phyla, but within species.[9] Louis Agassiz founded the Museum of Comparative Zoology at Harvard in 1859. The American Museum of Natural History was launched in New York ten years later, with support from Theodore Roosevelt and J. Pierpont Morgan. President Ulysses Grant laid the first stone.

As awareness of the diversity of living forms increased, there was a push to protect these organisms in their natural landscape, outside of museums. Yosemite, the nation's first park set aside for public use and preservation, had been planned with museums in mind. Frederick Law Olmsted, who drafted the original report on the valley and the giant sequoias of Mariposa Grove, justified protecting the area because of its value "as a museum of natural science" and because its plant life and natural beauty were at risk from the "bad taste" and "wanton destructiveness" of its visitors.[10] It was the landscape that mattered. Yellowstone National Park was created in 1872 to preserve its scenic wonders, with the panoramic canvases of the landscape painter Thomas Moran assisting in the campaign to establish the park. As in Yosemite, early advocates were concerned that commercial interests would exploit and destroy the beauty of the geysers and waterfalls. Almost as an afterthought, the two-million-acre set-aside would protect a species that had all but disappeared from the continent.[11]

The American buffalo had once grazed in Pennsylvania, New York, along the Appalachians down to the Florida Panhandle, and from the Allegheny Mountains to eastern California. There were probably 25 to 30 million bison in North America when Europeans arrived—a few million east of the Mississippi and west of the Rockies, the majority on the Great Plains.[12] William T. Hornaday described the sway American bison once held over the range: "They were so numerous they frequently stopped boats in the rivers, threatened to overwhelm travelers on the plains, and in later years derailed locomotives and cars, until railway engineers learned by experience the wisdom of stopping their trains whenever there were buffaloes crossing the track."[13] One settler reported that his Pennsylvania cabin had been demolished when a buffalo herd rubbed their backsides on its timbers. He and his companions killed more than six hundred in the following seasons. In 1825, the last bison east of the Mississippi, a mother and calf, had been taken in West Virginia.[14] By the 1870s, the buffalo were extirpated from almost their entire range. A few hundred hung on around Yellowstone, but even after the park was established, poaching continued; a local taxidermist paid up to $125 for each buffalo head.[15] The US Army was dispatched but had authority only to confiscate firearms and expel poachers from the park, not to prosecute them. Grant never signed a bill to stop the slaughter of bison in federal territories; another bill died in Congress in 1876, as did hundreds of thousands more buffalo.

In 1893, George Bird Grinnell, a well-known naturalist and editor of *Forest and Stream,* sent a writer to Yellowstone to report on the carnage. Accompanying an enlisted man on the trail of a notorious poacher, he came upon a camp where six buffalo heads had been bundled in gunnysacks and hoisted into a tree. The soldier followed the trails in the snow to the poacher, Ed Howell, who had just killed five more. The reporter's account of the daring capture and his outrage over the poacher's release at the order of the Secretary of the Interior soon reached Congress and the desk of John Lacey, chairman of the House Committee on Public Lands—a man who had been robbed by a poacher while traveling through the park a few years earlier. (That poacher had also been released by the army.) Lacey introduced the Yellowstone Park Protection Bill less than two weeks after Howell's capture. The law was a milestone, one of the first wildlife protection laws intended to *preserve* a species rather than extend the chance to hunt it. Convictions for poaching would now result in a $1,000 fine and two years in jail.[16] When Ed Howell was found in a barber's chair at the Mammoth Hotel in Yellowstone, he was arrested for returning after expulsion—the first person prosecuted under the new law.

Wildlife was under the gun: Winchester repeating rifles and Remington side-by-side shotguns were common sights in the West in the late nineteenth century (and later in Westerns). Large punt guns, fixed to their eponymous flat-bottomed boats, could kill dozens of waterfowl in a shot. Fifty-caliber single-shot rifles could take down the largest game. Wildlife harvests began to resemble the slaughter of men on the battlefields of the Civil War. Bounty hunters wiped out cougars, bears, and wolves from most of the East Coast by 1900. Market hunters took their aim at shorebirds, slaughtering golden plovers, buff-breasted sandpipers, long-billed curlews, willets, dowitchers, knots, godwits, and sanderlings. Once so abundant it was known as the prairie pigeon, the Eskimo curlew was dispatched by the wagonload on the Great Plains; hunters shooting more than they could carry left the bodies to rot in "piles as large as a couple of tons of coal."[17] The last one was shot in Barbados in 1963.

Other species survived, but in numbers so small they hardly made a difference. Attwater's prairie chickens, once "everywhere abundant" from the western Appalachians to Colorado and New Mexico, were killed by the carload—"squealers," the young of the year, were especially easy to shoot.[18] Almost all of their native grasslands have been plowed or paved. Members of the Class of '67, about 50 of the prairie chickens survive on the coastal prairies of Texas and 150 in zoos.

As wildlife numbers dwindled, there was a rise in field expeditions across the continent. Survey collectors—mostly hunters after skins and skeletons for museums—took advantage of new steamers, rail lines, mining camps, and tourist resorts that were opening up the interior of the country. In the 1890s, Vernon Bailey, a field naturalist in the federal Division of Economic Ornithology and Mammalogy, identified the biogeographical breaks between eastern and western faunas by stopping at railroad towns and collecting species as he made his way west. Such rail expeditions would continue to be mounted until suburbanization and the automobile destroyed the field in the 1920s.[19]

Even as they were busy collecting specimens, the scientists noticed a frightening decline in wildlife. Improved rails and refrigerated cars gave hunters easy access to big city markets. Sportsmen's groups, such as the Boone and Crockett Club (founded by Grinnell and Teddy Roosevelt) and the National Rifle Association had been pressing for protective laws for decades. Writing in *Forest and Stream*, Grinnell proposed the creation of an Audubon Society to protect wild birds and their eggs.[20] Leisure hunters and nature lovers formed the League of American Sportsmen in New York in 1898 "to stop market hunters from slaughtering game, and the killing of any innocent bird or animal, which is not game, in the name of sport or wantonness."[21]

Despite pressure from these groups, the public's attention was often focused elsewhere—on hostile Indian tribes or efforts to expand the nation to the Pacific Coast. Wildlife management and the regulation of hunting was left to the states. In monarchies, the ownership of game had been clear: hunting was a royal privilege and poaching was a crime against the king, punishable by fines, imprisonment, and even death. But the United States had no hard and fast rules until 1900, when John Lacey introduced a bill in the House of Representatives prohibiting the interstate trade of poached game and wild birds. Though he got help from Hornaday at the New York Zoological Society, Grinnell of *Forest and Stream,* and sportsmen across the country, Lacey's first attempt to call up the new bill failed when a New York congressman, representing the hat makers in his district, objected. Lacey, who detested the millinery industry almost as much as poachers, agreed to exclude "barnyard fowl" from the legislation.

The "Lacey Act" was revolutionary, essentially the first endangered species law. It enhanced existing legislation by making the possession, transport, or sale of wildlife a separate offense from the initial act of poaching. The act authorized the reintroduction of game and wild birds to areas where they were scarce or extinct. The import of harmful exotic species—specifically the mongoose, fruit bat, starling, and English sparrow, which Lacey viewed as "the rat of the air"—was prohibited. Even so, the law was too late for some. In March 1900, the same month that his measure was reported to the House, the last wild passenger pigeon was collected in Pike County, Ohio. McKinley signed the law a couple of months later.

A sometimes prickly alliance formed between the Audubon Societies, (predominantly urban, often female), sportsmen, and naturalists who worked in the field. For years, there had been conflict between conservationists and preservationists. Unabashedly utilitarian, conservation was the practical science of managing forests and game resources for sustainable use. Teddy Roosevelt and Gifford

Pinchot, the first chief of the US Forest Service, came to exemplify this approach, with Roosevelt asserting in a presidential address, "[T]he conservation of natural resources and their proper use constitute the fundamental problem which underlies almost every other problem of our National life." Planned, orderly development should replace "haphazard striving for immediate profit."[22]

John Muir, who founded the Sierra Club in 1892, championed the national park and the untouchable wilderness. A prodigious essayist and author of bestselling books, he fought to save the Hetch Hetchy Valley from being flooded to create a reservoir for San Francisco. The valley was lost, but he has come to be seen as a purist and prophet: "How narrow we selfish, conceited creatures are in our sympathies! How blind to the rights of the rest of creation!"[23]

Neither Muir nor Roosevelt was rigid in his beliefs. Muir tempered his biocentricity as he aged, even hiding it (as environmental historian Roderick Nash has noted) in his published writing and speeches "under a cover of anthropocentrism."[24] He knew that going before Congress to argue for national parks as places where beavers or snakes could exercise their natural rights would only bring ridicule. The older Muir insisted that nature was valuable for *people,* for protecting watersheds, for rest and recuperation, for aesthetic satisfaction. His remarks about the rights of nature remained in his private, unpublished journals until after his death.[25]

Roosevelt, though practical, cared about nature, too. In *A Book-Lover's Holiday in the Open,* he wrote, "Birds should be saved because of utilitarian reasons; and, moreover, they should be saved because of reasons unconnected with any return in dollars and cents. A grove of giant redwoods or sequoias should be kept just as we keep a great and beautiful cathedral. The extermination of the passenger pigeon meant that mankind was just so much poorer; exactly as in the case of the destruction of the cathedral at Rheims."[26] Rheims had been partially destroyed by German shellfire at the outset of World War I, two weeks after Martha, the last passenger pigeon, died in the Cincinnati Zoo on September 1, 1914.

The divisions between conservation and preservation were rarely straightforward. African colonial powers met in London in 1900 for what was perhaps the first international environmental meeting: the Convention for the Preservation of Animals, Birds, and Fish in Africa. At issue was the decline of game for colonial hunts. The leisure safari—Roosevelt's expeditions to East Africa, perhaps—was both an opportunity for conservationists to promote the preservation of large game and a disgrace: white men "hunted," Africans "poached."[27]

Meanwhile, North American bird populations plummeted from overhunting, and the feather trade moved offshore. Hornaday combined emotions and economics to appeal for increased protection not just at home but abroad. In Venezuela, he wrote, feathers were pulled from wounded birds, leaving them to die of starvation, "unable to respond to the cries of their young in the nests above." Injured egrets were tied down in marshes to attract other birds. They also attracted red ants, which ate out the eyes of the "helpless birds." Strong legislation, he insisted, would save the helpless, the farmers, and, well, everyone: "The good that would be accomplished, annually, by the . . . federal protection of all migratory birds is beyond computation; but it is my belief that within a very few years the increase in bird life would prevent what is now an annual loss [to insect pests] of $250,000,000. It is beyond the power of man to protect his crops and fruit and trees as the bird millions would protect them—if they were here as they were in 1870."[28]

When a bill to protect migratory game birds, supported by some hunters' groups, stalled in Congress, Hornaday commandeered the idea of protecting all migratory birds—songbirds had a growing constituency in the Audubon chapters—to broaden the legislation. The initial legislation was attached at the last minute to an agricultural appropriations bill to avoid an almost certain veto. On William Howard Taft's last day in office, March 4, 1913, he signed the farm bill, unaware of the rider prohibiting the sale of plumes, feathers, quills, and wings of wild birds.[29]

Five years later came the broader Migratory Bird Treaty Act, which protected birds that flew between the United States and Canada. More controversially, it protected all birds that traveled across state lines. Missouri challenged the law, posing the question: did the Constitution give the federal government the power to negotiate treaties and regulate hunting? In 1920, a Supreme Court decision affirmed the government's jurisdiction, and this principle has lead to several landmark pieces of legislation, including the 1937 treaty prohibiting the hunting of right and gray whales and the Bald Eagle Protection Act of 1940. These laws had a low cost to society—the species were relatively rare—and little opposition was raised. Later legislation was more comprehensive: the Wilderness Act of 1964, the Marine Mammal Protection Act of 1972, and the Fishery Conservation and Management Act of 1976.

Despite having the largest economy in the world, the United States has often been at the forefront of nature protection. Perhaps because we had such an early start in places such as Yellowstone and Yosemite, we hold a great deal of the world's protected landscapes. While the country comprises just 6 percent of the terrestrial world, 15 percent of the planet's protected areas lie within its boundaries. Thomas Vale, a geography professor at the University of Wisconsin, contends that Americans were the first society in the history of humanity to show deliberate and conscious restraint of economic development in an attempt to protect nature.[30]

The Endangered Species Act was part of this tradition, but it was also part of a deeper historical progression: expanding rights. According to Roderick Nash, the roots of this concept lie in the Magna Carta, which established political rights in England in 1215, and in John Locke's idea of natural rights—that every person shared a right to continue existing and all were equal before God and each other. The newly formed United States adopted these principles for its citizens with its Declaration of Independence in 1776.[31]

A pebble had been thrown into the pond; the ripples widened. The Emancipation Proclamation and the abolition of slavery in 1865 gave liberty to African Americans. In the twentieth century, rights were further extended to women (Nineteenth Amendment, 1920), Native Americans (Indian Citizenship Act, 1924), laborers (Fair Labor Standard Act, 1938), and African Americans (Civil Rights Act, 1957). As part of a wave as old as the republic itself, the Endangered Species Act (1973) gave nature, in all its forms, a right to exist.

5

The first species to be added to the original list were a bunch of kangaroos. No one was going to kick up a fuss over the listings, which did little more than prohibit the import of red and gray kangaroos from Australia. The head of the service at the time, an agency insider named Lynn Greenwalt, was determined to keep a low profile. In 1975, a handful of relatively uncontroversial species were put on the list, including the grizzly bears of the Lower 48 (dropped from the list in 1969 because they were common in Alaska); the American crocodile, long subject to conservation efforts; and three Hawaiian birds found mostly in parkland. The service was slow to stray beyond the big and the beautiful: by the end of the 1970s, only 39 invertebrates and 57 plants had been protected.[1]

The floodgates wouldn't hold for long. The listing of a three-inch fish called the snail darter went into effect on November 10, 1975, after Hill and Plater's lawsuit. It wasn't the first test of Section 11 of the Act, which allowed for citizen suits to protect species—the National Wildlife Federation had sought an injunction to stop an interchange on Interstate 10 that threatened the habitat of the Mississippi sandhill crane. The Fifth Circuit granted the injunction, noting that federal agencies had a "mandatory duty" to ensure that their activities didn't jeopardize protected species. The Department of Transportation then modified the project.

But the TVA couldn't move the Tellico Dam. The Authority was determined to overturn the decision of the Sixth Circuit, which had stopped the project. Attorney General Griffin Bell agreed to argue the case against the tiny darter before the Supreme Court. It was not unusual for the Justice Department to present the cases of federal agencies, but attorneys general rarely argued them. When they did, they never lost.

Early in the proceedings, Bell waved a jar of formaldehyde holding a dead darter in an attempt to dismiss the fish. Justice John Paul Stevens asked if it was alive. "I have been wondering what it's in if it is," Bell replied, in his deep Georgian accent. "But it seems to move around." As the laughter died down, Justice John Paul Stevens followed, "Mr. Attorney General, your exhibit makes me wonder. Does the Government take the position that some endangered species are entitled to more protection than others?" Bell's smile vanished.

When Zyg Plater, lead counsel for the darter, took his turn to rise for the respondents, the clerk of the court handed a copy of Exhibit 12 to each justice: a lithograph portrait of two snail darters, with big black eyes, wings as elegant as a Spanish fan, swimming over a pebbled riverbed. This was no pickle jar.

During oral arguments, Chief Justice Warren E. Burger accused Plater of using the endangered fish to stop the dam: "I'm sure that they just don't want this project," he said. "The snail darter was discovered, and became a handy handle to hold onto."[2] He was right, of course: It was much more than the fish. It was the valley's farmers, it was its fishermen, it was the descendants of the ancient Cherokee. It was history in the making.

After oral arguments, Bell may have won in the court of public opinion as he stood on the steps of the Supreme Court in a black morning coat, waving the dead darter in front of reporters. "This is what this case is all about," he said. "A silly little fish. It just doesn't make sense to me." Plater should have taken to those steps with the picture of the fish and a few farmers who stood to lose their land to the dam. But he still believes the artwork moved a justice or two.

Two months later, the Supreme Court rendered its verdict. In the majority opinion, Chief Justice Burger wrote: "It may seem curious to some that the survival of a relatively small number of three-inch fish among all the countless millions of species would require the permanent halting of a virtually completed dam for which Congress has expended more than $100 million. The paradox is not minimized by the fact that Congress continued to appropriate large sums of public money for the project, even after [being informed] of its apparent impact upon the survival of the snail darter. We conclude, however, that the explicit provisions of the Endangered Species Act require precisely that result."[3] The ruling upheld the precautionary principle—the burden of proof should fall on those who wish to take a potentially harmful environmental action. It was a permanent injunction against completion of the dam.

There were rumors that, because he wanted the winning opinion watered down, Burger switched sides after it became clear that the majority of justices was going to favor the fish. Carter Phillips, a clerk in the following term, said the chief justice routinely quoted the line about Congress being free to be stupid as long as it didn't violate the Constitution, "but he never discussed what went into the specifics of the drafting process." Stewart Jay, who clerked for Burger during the case, told me that the chief justice changed his vote "for entirely principled reasons."

"You'd think that was all you'd have to do," David Etnier, who had discovered the fish, mused, years after the case was decided. But Burger made clear that the case should now go back to Congress. "Everybody said he was begging Congress to bring sanity to this stupid situation," Plater told me. Even Talbot, one of the lead authors of the Act, was concerned that the darter was being used "by people who had different agendas entirely from the conservation of species." Justice Lewis Powell, in his dissent, was explicit: "I have little doubt that Congress will amend the Endangered Species Act to prevent the grave consequences made possible by today's decision."

Actually, the Senate Committee on the Environment and Public Works had already been working on just such an amendment even before the Supreme Court announced its decision. One of the bill's principal architects, Tennessee's Republican Senator Howard Baker, stressed the importance of balancing interests between development and protection. A project should continue, he argued, "if it is decided that the Federal activity is of more importance than the protection of a particular species." The bill's cosponsor, Democrat Jennings Randolph of West Virginia, added that environmentalists "armed with Section 7" (which required all federal agencies to conserve listed species) "will be able to literally shut down Federal construction programs by finding a remote species of mussel, snail, or fish at any project site."[4]

An Endangered Species Committee of cabinet members (including the secretaries of Agriculture, Army, and Interior, and the administrators of the EPA and the National Oceanographic and Atmospheric Administration) was established in 1978 to review conflicts between development and species protection. Concerned that the balance might tip toward the darter, Baker tried to make sure that appointees were friendly to the TVA and that the economist Charles Schultze, chair of the Council of Economic Advisors, was included. Soon nicknamed the God Squad, it had the power to grant exemptions to any of the Act's provisions. A population or species could be extinguished if it were in direct conflict with an administration's goals. The committee convened in January 1979 to review the dam. They were not impressed by the TVA's numbers. When Schultze, Baker's choice, proclaimed, "Here is a project that is 95% complete, and if one takes just the cost of finishing it against the benefits, and does it properly, it doesn't pay, which says something about the original design," the committee burst into laughter, along with the audience, and then voted unanimously: an exemption was not justified.[5]

Baker and Tennessee representative John Duncan were outraged, but President Jimmy Carter promised to veto any legislation that overturned the ruling. The TVA began to consider alternatives—something it had refused to do five years earlier—including removing the earthen dam and reopening the east channel of the river. Plater and his colleagues put forth a proposal to protect the Little T and establish a tourist route that ran the length of the river, from I-75 to the Great Smoky Mountain National Park, which they called the Cherokee Trail.

Meanwhile, Alfred Davis, whose family had worked the valley for generations, had been trying to meet with Senator Baker to plead the case for the families who would lose their land. Frustrated in their attempts to schedule time with the Tennessee politicians on Capitol Hill, he and some neighbors decided to drive to Washington.

"Somebody figured out that he was going to a meeting up in the Capitol Building," Davis recalled at a gathering years later. "So we split up and went to find the meeting. There was a whole group of us, but we'd broke up into twos, and we went and found this meeting. And there was this hallway lined up with some guards and beaucoups of reporters. And there were people lined up at least ten deep from us. And we started to weasel our way in to the front of this line. And just as we got to the front I'm thinking, How am I going to get to Baker? . . .

"Now we had on our best Sunday suits, mind you, but we probably still looked a little bit like hillbillies. And Wendell Perry's son from Vonore was with us; the young boy was about 16. When we got to the front there was a big guard. He looked over at us and said, 'Are you all reporters?' And before I could say anything, the young Perry boy said, 'No, we're just farmers.' And I thought, Well, we are going out the door now.

"But he turned around and said, 'Son, don't ever let me hear you say that again. The farmer is the most important person in the world. Farmers feed the world. What are you all doing up here?' And I said, 'We are from Tennessee, and we are against Tellico Dam, and we've come to see Baker, Senator Baker, and he don't want to see us.'

"The guard looks at us and says, 'The little weasel is right over our heads. He's got an office right up there.' And he said, 'He will be down here in just a few minutes, and if you want to see him, my suggestion is, go down that hallway, and there's an elevator. . . . Be standing in front of it. And when that door opens and when he starts to shut the door on you, you just step in the elevator with him.'

"So we quickly . . . got together when the door opened. . . . But he saw us, stepped out of the elevator, and guess what? He said he had heard we was in town and was so glad to see us!"

Happy or not, Baker and the rest of the Tennessee delegation were as determined as ever to finish the dam. In a near-empty chamber on June 18, 1979, Duncan added a rider to a $10.8 billion energy and water appropriations bill, without reading it aloud. No one objected. In less than a minute, the Endangered Species Act was amended. Most environmentalists read about it in the *Congressional Record* the following day. One didn't have to.

In 1977, Plater and Hank Hill had requested an economic study of the Tellico by the General Accounting Office, the investigative research branch of Congress known as the GAO, certain that it would expose the financial folly of the project. The chairman of the Merchant Marine and Fisheries Committee was New York Congressman Jack Murphy, whose assistant thought the study had merit. But, he told Plater, "You've got to give Jack a bribe. Everyone that gets something from Murphy has to give something."

He winked, went to a locker, and pulled out a giant jar holding a pickled sea snake, gift of an environmental group. "This is what he got for stopping a canal in Nicaragua. . . . They told him that it could come through his canal—if he authorized it—and swim all the way to Staten Island."

He turned to Plater, "So what do you have?"

Plater handed him Exhibit 12, the lithograph of the darter from the trial, now on sale for $18 a print.

He looked at it for a while and said, "I guess that will be enough."

Now all they needed was a Tennessee congressman. A freshman named Albert Gore, Jr., heard them out and said they had his complete support. But later, when Hank Hill arrived at his office, he found Gore's receptionist on the phone, asking a colleague to ignore Hill and Plater's requests to meet. When she hung up the phone, she asked, "Can I help you?"

"We're the people that you were just talking about."

Though he wouldn't put his request in writing, Gore agreed to give his oral support to Hill. Soon after Hill left, the congressman changed his mind. Murphy's staffer gave Plater five minutes to reach the chairman before she would cancel Gore's request. Plater got to the committee room in three. When Gore reached Murphy to rescind the letter, he was told, "Those boys beat you fair and square, Al."[6]

Tennessee Senator James Sasser soon wrote a letter to the GAO supporting the study. Although the GAO study didn't take a position for or against Tellico, it was remarkably frank: "The Congress should prohibit by law the Authority from spending any more appropriations for work on the project that would further endanger 'the darter' or be wasted if the dam was not completed, pending an intensive economic rethinking of the project."[7] The dam's ardent supporters had no mind for rethinking.

The rider went through on a voice vote in a near empty House, with no objections. Later that summer, an attempt to strip it from the bill lost by a hundred votes: Gore supported the dam; the freshman Newt Gingrich supported the fish. Senate minority leader Baker proposed similar legislation that summer. It failed to pass, but he reintroduced it after summer recess with a now famous flourish: "The Tennessee snail darter, the bane of my existence, the nemesis of my golden years, the bold perverter of the Endangered Species Act, is back. He is still insisting that the Tellico Dam on the Little Tennessee River, a dam that is now 99% complete, be destroyed. In the midst of a national energy crisis, the snail darter demands that we scuttle a project that would produce 200 million kilowatt hours of hydroelectric power and save an estimated 15 million gallons of oil." Baker didn't mention the farmers who would lose their land or the Cherokee who would lose their sacred sites. He got the votes he needed, 48–44.

Was there still hope for the darter? At the urging of Andrus and many environmental groups, the president had decided to veto the bill. But then, on the night of September 25, 1979, Plater received a call patched in from Air Force One. President Carter "clearly wanted my forgiveness, my absolution," Plater told Trout Unlimited.[8] He signed the bill the following day. "It came so close," Plater told me. "If Jimmy Carter had said, 'Look you've all heard about how this is a silly little fish stopping a huge hydroelectric dam. Let me tell you about the farmers, let me tell you about the river. Let me tell you about the real merits of that little dam,' he might have saved it all."

On November 13, 1979, armed federal marshals were sent into the valley to remove Thomas Beryl Moser and Nellie McCall from their land. Bulldozers destroyed homes before the farmers' eyes. At 75, now widowed, McCall said she didn't know where she would move now that she had lost "the 91-acre farm her husband had purchased in 1939."

On November 29, TVA engineers severed the steel cables that had hoisted the floodgates above the small dam. The water began to rise. It settled the shoals. It shrouded the sacred sites of the Cherokee. It wicked up through the abandoned croplands. It took about a month to fill the reservoir. The Little Tennessee, which had flowed for some 200 million years, became a lake.

From an embayment in the river, more than 590 darters were captured and transplanted. Death was probably slow for the several thousand that remained. Perhaps some swam upstream in search of new shoals, shoals they would never find. Some could have remained in place, living out their final years in the flat water. But without the riffles they needed to reproduce, all that were left were the senescent. Eventually even they died. The snail darter was never seen on the Little Tennessee again.

Broad political support for the Endangered Species Act was beginning to founder as the economic and political realities of saving every species in the United States set in. In its original form, Section 7 was 129 words long. It was blunt and direct, requiring federal agencies to ensure that their actions "do not jeopardize the continued existence" of listed species. In 1978, Congress and the Interior Department tempered the Act. The new consultation procedures were more expansive—the section now runs to more than 4,600 words—and more complicated. "Do not" was softened to "are not likely to."

The darter survives. Etnier had been right: there was a native population in the nearby Hiwassee River, a relatively clear stream just to the south of the Little Tennessee. Many of the darters collected by the TVA from the Little T were translocated there. The darter also showed up in some surprising locations. Shortly after the dam was closed, Etnier and his colleagues seined a stretch of the Chickamauga Creek, a rather polluted waterway that runs through Chattanooga. "Well, I'll be a son of a bitch," he recalled saying, when they netted six snail darters. "We kept one and let the others go." Between Etnier, the TVA, and the feds, darters were found in several streams along the main branch of the Tennessee as far south as Alabama.

"The real surprise came when we found one in Little River," Etnier said, when I met up with him on the shores of Tellico Lake, at what he described as "a wake for a fish," the thirtieth anniversary of the lowering of the floodgates. The spring where he had captured his first darter was completely inundated, the farmer's land along the spring subdivided into dozens of high-end homes. "My students had been doing masters' theses and dissertations for many, many years there, and we thought we knew just about every individual fish by name in the river. One of my students went up there one day, and he caught a snail darter. We couldn't figure out what the heck had happened. Next thing, we caught one on the Holston River right under the Interstate 40 Bridge in Knoxville. Then we started figuring it out. We had had a few snail darters left over after the dam was closed. In the recovery team meeting, we decided that we'd already put enough in the Hiwassee. There weren't many other places left to put them.

"Somebody said, 'Let's put them in the Lower Holston.'"

"Might as well just flush them down the damned toilet as put them there," Wayne Starnes, who had studied with Etnier in Tennessee, replied.

The waters around Knoxville were polluted, giving the river bad dissolved oxygen problems—the tailwaters from many of the dams had become biological deserts. But then the TVA started pumping oxygen from tanker trailers into waters just above the dam. "All of the sudden things started showing up that we didn't even know were in the streams," Etnier told me. "People started transplanting snails and clams. They did fine, and the fishing got good. Just wonderful, wonderful places. It's certain to us now that the snail darters were just barely able to hang on before the oxygen problems were solved. They had a head start on the other fish. It was a great new world to live in. And they exploded." One darter even showed up in a stream above the Little T. The snail darter was downlisted to threatened in 1984.

New species were soon added to the endangered species list at a rapid clip: 159 in 1976 alone, including 24 mussels in the United States and more than a hundred foreign vertebrates, such as the Chinese alligator and Przewalski's horse. The first plant species were listed as endangered in 1977: four endemics on San Clemente Island. After a few new hires, the Endangered Species Office in DC was starting to look like a decent university wildlife department—there were mammalogists, invertebrate biologists, botanists, an ichthyologist, and a herpetologist. They initiated listings for more than three thousand species, though most of the proposals were eventually dropped.

That many biologists were bound to get themselves into trouble sooner or later. If you've ever been to an Ichs and Herps meeting, you know it was going to be the herpetologist who got there first.

During the Carter Administration, Ken Dodd, who had moved up from Mississippi State to join the office, found himself watching a "fluff piece" on the evening news about a fancy new restaurant that was serving rattlesnake sautéed in red wine, with string beans and red cabbage on the side, at $9.25 a plate. "I didn't give a shit about that," he told me, "but they said it was coming from Pennsylvania." Dodd had just read a scientific paper on the decline of the Pennsylvania rattler, now so rare that it was protected by state law. "I thought a nice friendly letter would alleviate the problem."

Dodd sent a note to the restaurant's owner, Dominique D'Ermo, advising him that selling a protected species across state lines could be a violation of federal laws and "respectfully requesting" that he find another source for snake. D'Ermo was well connected: his French restaurant was Interior Secretary Cecil Andrus's favorite in DC. Andrus apologized to the chef and sent a letter assuring him the snake was not endangered. "I hate rattlesnakes," he told the press. A few days later, Dodd was fired for using official Fish and Wildlife Service stationery to express his concern, an act that usually rates just a $300 fine.

But Dodd was connected too—the letter of dismissal was on the front page of the *Washington Post* the day after he received it. Fifteen environmental groups came to his aid. Congress threatened to open an inquiry into the inappropriate firing of a civil servant. The secretary backed down—after assuring the press that rattlesnake would still be served at Dominique's. The chef assured reporters that from then on, all his snakes would be from Texas—where they were common, fat, and tasty.[9]

Two years later, Ronald Reagan shut the listing program down. According to Don Barry, who was the head attorney for the Fish and Wildlife Service at the time, Interior Secretary James Watt rolled into town, "with a little blueprint . . . prepared by the Heritage Foundation," a conservative think tank. "They had what they called their Dirty Dozen legal opinions that needed to be overturned. . . . Most of them I had written on the ESA." Not a single species was added during the first 376 days of Watt's tenure. Controversial petitions were ignored if they interfered with federal projects. According to Jim Williams, who was a zoologist in the Endangered Species Office,

"Watt's marching orders were to add nothing to the list until you get all the others recovered."

Even when biologists finished a listing package, the lawyers would slow things down. "You'd send it over to the solicitor's office," Williams told me, "and they would sit on it for probably a couple of months. You'd keep bugging them, and maybe go downtown and say, 'Hey, you know I sent that over to you a couple of months ago—have you had a chance to look it over?' 'Well, no, but I'll get to it next week.'

"You'd keep bugging them, and finally they'd find all these things were wrong with the listing and send it back over for revision. This went on and on. It was a game. They didn't come right out and say we're not going to do anything, but that was the sum total of it."

Watt's interests in diversity appeared to lie elsewhere. "We have every kind of mix you can have," he boasted at a lobbyists' breakfast of a commission he had appointed. "I have a black. I have a woman, two Jews, and a cripple. And we have talent."[10] His resignation followed soon after.

Partly as a response to the early Reagan years, Congress amended the Endangered Species Act in 1982. Timely responses to listing proposals were now required. Decisions had to be made based on the best scientific and commercial information available, not on the economic reviews the administration had been using to reject listing packages. The average listing rate rose to 40 species per year.[11]

But having biologists inside the Beltway remained a problem for the administration. "They found they couldn't control us," Williams said. When he was told that the department was going to reject a petition to list the Atlantic salmon, he said, "I'm not putting my name on a document in the *Federal Register* that says this shouldn't be done. It's wrong.

"That sort of thing just drove them up the wall. They were so used to saying 'Do this,' and we'll just go away and do it. Never ask questions." The biologists had good connections with the press and national environmental groups. "So eventually they said, 'Okay, we're going to send you guys out to the hinterlands.'" The Reagan administration began to dismantle the Endangered Species Office in DC. Biologists have been working from regional offices ever since.

Oh, and about that dam? It never produced the benefits that the TVA claimed it would. Tellico generated a tiny bit of power by diverting water to the Fort Loudoun Dam, but the industrial development didn't happen. When *The CBS Evening News* covered the story five years later, reporter Bernard Goldberg noted that the model city promised by the TVA had never appeared. So why not a toxic waste repository for this once beautiful valley? the Authority wondered as it considered alternatives. The leak of its secret memo to the press got that plan squelched.

Thirty years later, farmers who refused to sell to the TVA still remembered the federal marshals showing up on their land to escort them off. In their place, an upscale golf and lakeside development, once partly owned by Wal-Mart's Sam Walton, has more than thirty thousand residences, most of them bought by second-home owners and retirees. "The farmers," Plater has written, "were unable to repurchase their condemned lands and now can go onto their old properties only in the capacity of servants or employees."[12]

In one respect Tellico was a success. Building a dam in the United States would never be easy again. According to David Conrad, a water resource specialist at the National Wildlife Federation who was involved in the case, the fight "kind of broke the back of the TVA. They were never the same." Congress stopped rubber-stamping projects. Dams even started coming down. "The snail darter was an all-consuming part of my life for several years," Jim Williams, who put the listing package together, said, "and it's where I discovered that the intersection of science, politics, and conservation can be dangerous, difficult, and impossible to manage and nearly impossible to win. . . . We lost the snail darter battle, but I think we won the war." That's not something you hear a conservation biologist say every day.

The TVA had been working on an impoundment on the Duck River of Tennessee, which has more than 50 mussel species and 150 fish, since 1973. Then they learned that two of these mussels were endangered and would be jeopardized by the project. Williams recalled, "They immediately said to themselves, 'Hmmm, do we want another one of those? No, we don't want to do that again. Stop the dam.' About ten years ago they went in and blasted it out. I have a chunk of rock as a souvenir on my desk.

"Ah," Williams sighed. "A little piece of the Columbia Dam." I imagined the satisfaction of an East Berliner coddling a piece of the Wall. Except for a small dam at its headwaters, the Duck continues to flow unimpounded.

Endangered species have become the flagships, the surrogates, for their ecosystems: Chief Justice Burger's "handy handle." The red-cockaded woodpecker helped fight the crime of destroying the last of the longleafs. In the Pacific Northwest, the northern spotted owl carried the weight of the old growth forests on its white-barred wings, scapegoat of a dying industry. It was a bitter fight: soon after it was listed in 1990, a spotted owl was nailed to a sign outside Olympic National Park, a match sticking out from its breast. "If you think your parks and wilderness don't have enough of these suckers, plant this one," read a typewritten note. The controversy marked the end of the unbridled exploitation of ancient forests in the United States. Timber harvest on the 24 million acres of federal land within the spotted owl's range has declined by more than 90 percent since its peak.[13]

The battle between private interests and public benefits had long gone to the former. But in the effort to save species, we recognized the need for limits, the need to balance ecological sustainability with economic growth—with dams, highways, and even housing developments. The Endangered Species Act seemed like the most altruistic of laws. By legislating the protection of rare animals and plants regardless of cost, the Act uncoupled the fate of species from the free market. In protecting them, there was also a chance that we would save ourselves.

6

In his treatise on self-regulation in animals and plants, Carl Linnaeus praised the economy of nature. As he saw it, the epitome of wealth, at the root of the science of economics, was the domestication of new plants, such as foreign spices and teas. David Hume, a contemporary of Linnaeus, regarded economic processes as part of nature. Goethe claimed that "nature was the perfect economy." It was a common seventeenth- and eighteenth-century view: economic relationships were seen as a reflection of the natural world. Even Adam Smith had a keen and long-standing interest in natural history.

"Wealth was equated with the fruits of the earth and sea," Margaret Schabas writes in her fascinating *The Natural Origins of Economics*.[1] It was minerals, of course, gold and silver, but also fish and exotic plants. "[A]n economist without knowledge about nature," Linnaeus wrote, was "like a physicist without knowledge of mathematics."[2] He saw to it that chairs in economics were established in every university in his native Sweden.

Oeconomia, the art of household management, is as old as Homer, or at least Xenophon, who wrote a book on it sometime around 400 B.C., but for thinkers of the Enlightenment, human reason was a derivative of natural instincts. Nature was wise, just, and benevolent in creating wealth, "the bustle of the world," for naught. Adam Smith realized that the man who worked incessantly, serving "those whom he hates," would begin to find "in the last dregs of life, his body wasted with toil and diseases, his mind galled and ruffled by the memory of a thousand injuries and disappointments . . . that wealth and greatness are mere trinkets of frivolous utility"; what mattered was "ease of body or tranquility of mind."[3] As Schabas notes, the whole goal of wealth was "to restore the abundance and complete leisure of the Garden of Eden."[4]

Such ideals began to lose ground in the mid-nineteenth century, most famously through the work of John Stuart Mill, who supported women's rights, opposed slavery, and lamented a world that was in danger of losing its wild animals and plants. But he could see nature as imprudent, unjust, and cruel, proposing that human economy was separate from the natural order. The bridges, wells, and lightning rods that people constructed were acknowledgments that "the ways of Nature are to be conquered, not obeyed."[5] To many Victorians, the economy became a product and symbol of human deliberation, divorced from nature, *denaturalized.* At the same time, solar energy, which had driven preindustrial societies by producing food, fiber, and fuel, began to be supplemented and replaced by fossil fuels—first coal, then oil and gas. Ecology sat on the bench for more than a century, while economists battled on the playing field.

In the 1970s, a few rogues emerged. Economist Nicholas Georgescu-Roegen modeled the economy as a living system: it consumed plants, animals, ore, and the fossilized remnants of long-dead living creatures. It spewed out waste: heat, carbon dioxide, old clothes, and now plastic packaging, cell phones, and computers. Economic growth could not override the laws of thermodynamics. One more Cadillac, the luxury item of Georgescu-Roegen's time, was one less plow or spade for a future generation; "biology," he wrote, "not mechanics, is the true Mecca of the economist."[6] Georgescu-Roegen's work was influential among a few, but remained largely overlooked in textbooks and economics departments.

And then one of his students, Herman Daly, became a senior economist at the World Bank in the 1980s. There, he drew a big circle around the economy and labeled it *ecosystem* to make it clear that the inputs taken from the natural world became the outputs returned to it as pollution. Daly called this early attempt to get beyond the idea of endless economic growth "steady state." In short: don't use resources faster than they can be replenished; don't emit waste faster than it can be absorbed. The World Bank wasn't ready for it. Even though the 1992 report he was working on was dedicated to sustainable development, his diagram was dropped from it.[7]

For ecological economics, 1997 was a watershed. Bob Costanza, one of Daly's collaborators, met with colleagues at the National Center for Ecological Analysis and Synthesis in Santa Barbara to undertake a gargantuan task: calculate the value of all the services provided by all the ecosystems, from the forest to the floodplains to the open ocean, across the world. In a paper published in *Nature,* they came up with a number—$33 trillion a year—a headline-grabbing figure that was almost twice the global gross national product. Arguments ensued over the methods of extrapolation and the neglect of services provided by human-dominated landscapes, but few in the field dismissed the idea entirely. Gretchen Daily at Stanford edited a book called *Nature's Services: Societal Dependence on Natural Ecosystems,* setting forth a framework for inventorying and valuing these services. Both of these works would prove enormously influential.

What is the value of a wild bee, a wetland, a bat that eats pests or pollinates desert cactuses? A few numbers: The benefit of coastal wetlands for hurricane protection has been estimated at $8,240 per hectare per year; in the United States, coastal wetlands currently provide storm protection services valued at $23.3 billion annually.[8] Studies from Portland, Oregon, to Portland, Maine, have found that every dollar invested in protecting watersheds can save up to $200 in costs for new water treatment and filtration facilities.[9] The services provided by coral reefs around the world include coastal protection, fisheries, tourism, and biodiversity: they have been valued at up to $172 billion a year.[10]

Around three-quarters of the world's 240,000 species of flowering plants—including most of our crops—rely on pollinators to reproduce. A 1998 study estimated that the economic value of pollinators for the most common crops in the United States was $40 billion a year.[11] But many are now in trouble. The recent collapse of honeybee colonies imported from Europe has led to a closer look at our native bees. The western bumblebee, *Bombus occidentalis,* once the most common bee in the West, has disappeared from areas around San Francisco and is in catastrophic decline throughout most of its range. *Bombus affinis,* a bumblebee once abundant on the East Coast, has almost entirely vanished.[12] In the UK, the native black bee, *Apis mellifera mellifera,* was cast out in the nineteenth century in favor of highly productive Italian bees. But after the crash of imported bees—more than 30 percent were lost in the winter of 2007–2008—researchers and breeders have begun looking for alternatives. The black bees, whose honey graced the tables of medieval kings, may be called upon to pollinate the island's crops once again.[13]

Just as the unification of Darwin's and Mendel's theories in the 1950s gave rise to the field of biological evolution, so ecological economics attempted to synthesize human activity and the natural world. Any beginning ecology course will tell you that growth is density-dependent; an overuse of resources causes a population to crash. Did that mean endless growth was impossible? Steady-state economics held that it was. Some called it "true-cost economics," where every dollar comes at the price of a tree cut down, the water or air polluted, a species lost to a city built atop its entire range.

In the eyes of these economists, nature is once again benevolent: it is the provider of goods, food and fiber, and services—cycling nutrients, forming soils, supplying water, cleaning the air, controlling erosion, and capturing carbon. It protects us from catastrophes: hurricanes, droughts, and floods. Natural fire regimes provide a barrier against wildfires spreading into towns, taking out the duff and the hot-burning hardwoods. Wetlands form vast natural levees, absorbing storm surges and protecting against floods. When we lose this ecological structure, we lose function—including those ecological services we need to survive.

When traditional, or neoclassical, economics began, the forests were plentiful, the oceans full of fish. Economists could ignore natural resources, and they assumed that people were rational, selfish, and insatiable. If there weren't enough mills or boats, that was a question to be answered with capital, labor, and technology. Things began to change as human impact spread around the world. By the late twentieth century, as Josh Farley, an economist at the University of Vermont, puts it, "Things created by humans are abundant. Things created by nature are scarce. The most critical scarcities are probably those services provided by nature like climate stability, water purification, and the provision of food, fiber, and fuel." If we want to protect nature and its services, the ecological economists believe, then we need to achieve ecological sustainability and a just distribution of resources. "If people don't have enough to feed their kids," Farley said, "they don't care about the future."

As the New England snow began to brighten to the fluorescent blue of winter dawn, the desk lamp flooding my office with light brought out the cluster flies. Making their way through the cracks and windows in my leaky nineteenth-century farmhouse, they swarmed the lamp, clung to ceiling and walls, and one even buzzed a cluster-fly backstroke across my cup of coffee. Too few stuck to the flypaper attached to the window.

Outside, a chickadee arrived, perching on the telephone wire. It was so much more than a black-capped beauty. Every morning the songbird performed an ecosystem service, taking the flies from the eaves, pecking at the glass and the sill, eagerly sticking its beak into the crevice between the chimney and the wall. I had felt a tiny bit of guilt over the flycatchers I'd hung—even as I found myself repelled by the ever-darkening clutch of hairy black bodies, eyes, legs, and wings stuck to them—but my conscience was eased when I learned that cluster flies had been introduced from Europe about 150 years ago. It had taken them a while to become established, because they deposit their eggs in earthworms—and the flies had to wait for the worms to arrive from Europe first.

By midmorning, the chickadee retreated to the sumac. A harrier cruised the corral in search of a careless rodent. People have long understood the value of natural systems. Farmers depended on pollinators, ranchers on grasslands and hay. When the Mormons arrived in Salt Lake Valley in 1848, a late frost and a spring drought destroyed most of their crops. Hordes of flightless katydids, later known as Mormon crickets, moved down from the surrounding mountains to consume what was left in the fields. But then, it is said, flocks of gulls flew in from Great Salt Lake. The settlers believed the birds, in devouring the crickets, had saved farms and their lives. The first monument erected to commemorate the service of wild birds stands in front of the Salt Lake Assembly Hall in Utah: two bronze versions of the Utah state bird, wings held high, atop a granite sphere. It remains one of the few public memorials to commemorate an ecosystem service provided by a wild animal.

Simply by protecting species and their ecosystems, we get so many benefits for free. Call them *nature's services, natural capital, ecosystem services,* even *ecoservices,* if you will—economists, ecologists, and conservation biologists have been trying these terms out for years to see what sticks with the public. Whatever it's called, this capital can be cultural: a bison on an old Sioux hunting ground, a bald eagle flying over the Potomac, a wolf pack in Yellowstone. It can be recreational, ranging from whale watching, to a trip to the Okefenokee, to a hike out your back door. It can even save lives—ecosystems provided a new drug for breast cancer, thanks to the Pacific yew—and restructure the tree of life. A heat-loving bacterium, *Thermus aquaticus,* catalyzed the molecular revolution and helped decode the human genome.

But why do we need biological diversity? If one species disappears or another is introduced, how will that make a difference to our well-being? Ecologists have developed several powerful metaphors and hypotheses.

Portfolio Theory. A portfolio containing stock in only one company, or holding nothing but stocks, bears a greater risk of losing value as markets fluctuate. Similarly, an ecosystem with only a few species is exposed to local and global change. In the financial world, investors minimize volatility by keeping savings in a diversity of stocks, bonds, and cash; in the same way, a greater variety of genes, species, and natural communities can stabilize an ecosystem, reducing the risk of losing natural capital after a disturbance. A diverse ecosystem is resilient and able to absorb the loss of a single species without a catastrophic shift. Endangered species conservation is a long-term investment; in turbulent times, we should hold on to some "cash" (captive populations) to be "invested" (released) when the habitat has been protected and the regime made more stable.

The Rivet Hypothesis. In 1953, Aldo Leopold wrote the oft-quoted: "To keep every cog and wheel is the first precaution of intelligent tinkering."[14] The loss of any species can be crucial to the performance of the entire ecosystem. For several decades, ecologists nodded their heads in agreement over this, but no one experimentally tested the idea that losing species could change ecosystem function until the early 1990s, when Shahid Naeem and colleagues at Imperial College London developed a machine they called the Ecotron. Sixteen growth chambers, two meters to a side, controlled humidity, temperature, the generation of rain and wind, and the shift to red light at dawn and dusk. Each microcosm was relatively simple, a meadow in miniature. The most diverse were comprised of thirty species of plants and animals—such as earthworms, herbivores, and predatory insects—and an inoculate of microbes from nearby meadow mud. The lower chambers had subsets of sixteen or ten species each. Although these chambers contained only a few species and a few subsets, the results were convincing: the amount of carbon dioxide absorbed by the communities, the biomass they produced, the fertility of the soil, and the amount of water retained by the ecosystems declined with the loss of species. "Because everything was held constant among the ecosystems except for their biodiversity," Naeem wrote, "the only conclusion we could come to was that our monkeying with the number of species was sufficient to drastically change the way the ecosystem functioned."[15]

The loss of biodiversity caused a decline in services in the field, not just in the Ecotron. At the University of Minnesota, David Tilman has done extensive, large-scale work on 38 acres of Minnesota prairie, manipulating up to 32 species of grassland plants. The first species lost had only a small effect on primary productivity, but the decline steepened as diversity decreased.[16] Higher diversity resulted in more biomass, more production, increased retention of nutrients, greater resistance to invasive species and plant pathogens, and greater stability. More diverse communities even consumed more carbon dioxide. Once too many species dropped out, the system no longer functioned fully or crashed completely.

How does this work? There are at least two factors involved. One is simple probability. The more species you remove, the greater the chances that an extraordinarily important one will be lost (the fatal rivet removed from a plane). The other is *complementarity,* which strikes at the heart of Leopold's quote: the more species you have, the more ways they make use of limited resources. Variations in light, water, nutrients, and space affect organisms differently. The supply of food, biofuels, and other ecosystem services to humans are enhanced by this variety. All aspects of our well-being have their roots in biodiversity. Lose the web, lose them all.

How much is a single species worth? What about an acre of wetlands? Economists break *worth* down into two types: (1) *use,* or market, values, and (2) *nonuse* values, which are often existential or spiritual. Market values can be direct: the timber we use to build our homes, the fish we pull from the ocean: they're *goods.* Salmon fishing in the Pacific Northwest supports 60,000 jobs and more than $1 billion in personal income; marine recreational fishing generates about 350,000 jobs and $30.5 billion a year.[17] How much is an old-growth sequoia forest worth in board feet? A school of bluefin tuna in kilograms of sashimi? A whale in *kujira,* or whale meat?

Whale meat? Old-growth sequoia? Our perception of many species has shifted over the years: where once we saw them as goods to be extracted from the forest or the sea, now we value them for the services they provide. Many old whaling ports and timber towns traffic in tourists these days. Even the city of Forks on the Olympic Peninsula—the former, self-proclaimed "Logging Capital of the World," once ground zero for the spotted owl controversy—is cashing in on hikers, hunters, and, yep, birders.[18] Views on the giant bluefin lag behind. It is still mostly valued, at the end of the line, as sashimi.

One of the most straightforward ways to measure such recreational services is through travel costs. How much will people pay to see a bald eagle or a manatee? Whale watching in the United States is an industry worth $956 million a year; there were nearly five million whale watchers in 2008, increasing the diversity of jobs in areas suffering from the collapse of fisheries.[19] It is not out of the question that someday the International Whaling Commission, started in the 1940s to save the whaling industry from hunting itself into extinction, will become a whale-watching commission, overseeing the impacts not of harpooners but of *ooh*-ers.

Another way to measure services is through avoided costs. Each spring, more than one hundred million Brazilian free-tailed bats disperse from caves and highway underpasses in the southwestern United States to feed on insects (including the fall armyworm, cabbage looper, and cotton bollworm) and other agricultural pests. Cutler Cleveland of Boston University and his colleagues estimated that these bats reduced crop damage and eliminated the need for early applications of pesticides in the cotton fields southwest of San Antonio, where more than 1.5 million of them provide about $741,000 of these services to local communities.[20]

The advantage of studies like these is that they can help change the all too commonly held perception that wildlife protection is an economic burden. There is plenty of evidence that the creation and restoration of healthy ecosystems generates ecological and economic benefits: sustainable fishing on Philippine reefs provides benefits of coastal protection and tourism that greatly outweigh destructive dynamite fishing; small-scale farming and low-impact logging in Cameroon has a higher economic value than widespread clearing for plantations. (Perverse government incentives, via tax breaks and subsidies, can override these values.) For every dollar invested in preserving habitats and populations, a hundred dollars of ecological services are provided. A global network of nature reserves covering about 15 percent of the world could deliver $4–$5 trillion in goods and services each year.[21]

The battle over the northern spotted owl and old-growth forest in the Pacific Northwest was particularly bitter, pitched as a choice between owls and jobs. Subsequent economic analyses showed no effect on employment in the timber industry after the owl was listed, even in Washington and Oregon, where the battle was fought. The steepest decline in employment had come well before environmental laws were put in place, back between World War II and 1964, when a third of Oregon's large mills had closed. Eighty-five percent of the smaller sawmills (the focus of the debate) had shut their doors before the owl was listed. After the Wilderness Act was passed in 1964, employment actually increased. It wasn't environmental laws that doomed the logger but a lack of forest management—by the late 1980s, almost all of the vast tracts of old-growth trees were gone.[22]

What are the future benefits of biodiversity for everyone? In his class on environmental economics, Robert Stavins at Harvard's Kennedy School of Government likes to use the example of an elderly woman residing in Cambridge. There's a small community playground nearby; she might be willing to pay five dollars for her grandchildren to use it—that's a *use value*. She might not have any grandkids, but would be willing to pay five dollars in case her daughter has children; this is an *option value*, bequeathed to future generations: a willingness to pay a certain amount today for the future use of a resource. Or she may not have any children or grandchildren at all; no one in her family will ever use the playground and, besides, it's across town—but she is willing to pay five dollars to keep it around, just to know it is there: this is known as *existence value*. And let's say the woman is also interested in tigers. Though she knows she'll never visit the forests of India, and that she'll never see a wild tiger, she is willing to pay to keep them around.

Now we're approaching the realm of neuroeconomics and *biophilia*, the hypothesis that our evolutionary history has blessed us with an affinity for other beings by descent proposed by Harvard biologist E. O. Wilson. How much does your happiness increase when you're in the woods? Or when you're stuck in one of the biggest cities in the world—how much would you pay for a view of the oaks and elms of Central Park? What *is* the value of a humpback whale on Stellwagen Bank, a gorilla in the Rift Valley of Africa, or a blind salamander in a Texas cave?

A 2008 review by Berta Martín-López and her colleagues at the Universidad Autónoma de Madrid has provided a few monetary values. Species that were hunted, fished, or considered a tourist resource were often valued the most highly: red deer ($207), moose ($145), humpbacks ($128), bald eagles ($115), and chinook salmon ($127). (Rarity and size also played a role; we're willing to pay more for threatened species if they're big.) The harder-to-love species were worth far less: the striped shiner ($6.83); the coyote, that livestock predator ($5.49); the tiny Eurasian red squirrel ($2.87).[23]

Nonuse existence values represent, in a sense, the satisfaction that people all over the world would realize in bringing the extinction rate down to zero. Economists sometimes talk as if such values were absolute. But visitors to Grizedale Forest in Britain, when asked how much they would pay for a woodland conservation scheme, were willing to pay almost twice as much on the days the interviewer wore a suit and tie than when he wore a T-shirt and shorts.[24]

Existence values may seem pure—or, as Bryan Norton, who teaches philosophy at Georgia Tech put it, "Species that have only existence value have, by hypothesis, no use"—yet estimating them has limitations beyond the dress code. Norton argues that willingness to pay cannot get at the intrinsic value of a species. Say, for example, there are two hostages being held for ransom: one good-looking, friendly, from a wealthy family; the other ugly, cantankerous, and alone. Just because the first hostage can attract a huge ransom while the second gets none does not mean the first has a higher intrinsic value; most ethicists would agree that the intrinsic value of a human is not determined by his power to attract. Similarly, a willingness to pay cannot determine the intrinsic value of a species, whether it's a humpback whale or a dung beetle. Even putting a fixed number on an endangered species risks ignoring its transformative value—its ability to change our worldview.[25]

All this talk of cash starts to ruffle feathers. "What about their spiritual value?" one conservation biologist challenged, as I followed him into a pond. He and his students were dragging for plankton to spike a cattle tank filled with endangered Mississippi gopher frogs. "We have an ethical and moral responsibility to protect them. Our lives will be poorer if we don't have this species around." If every species has to pay humanity its dues, argues one environmental ethicist, then nature conservation becomes extortion, a protection racket.[26] "What can this centipede do for me?" as Stephen Meyer of MIT has put it.[27]

There are those who argue on principle—a Thoreauvian ideal—that assigning values to nature and its services will play into the hands of industry and bleach the "conservation lexicon bone white."[28] Or as Thoreau wrote, in celebrating the "rich and various crop . . . unreaped by man": "There are berries which men do not use like choke berries—which here in Hubbards Swamp grow in great profusion & blacken the bushes. How much richer we feel for this unused abundance & superfluity. Nature would not appear so rich—the profusion so rich if we knew a use for everything."[29]

Perhaps money isn't particularly satisfying in determining our values. It isn't a great metric, for example, in judging the value of a human life: How much would you be willing to pay for the existence of your firstborn? Would it depend on what she did for a living? Should the family of a stockbroker be paid more than the widow of the fireman who died trying to rescue him? Kenneth Feinberg, in charge of the Victim Compensation Fund after 9/11, was required by law to calculate the financial worth of each victim. He was screamed at, spat on, and sued by some of the families of the highest earners. After the fund was distributed, he made the case that any future compensation should be equal for everyone, regardless of income.

Environmental philosopher J. Baird Callicott suggests that our values might be better reflected in the penalties for violating laws.[30] Sentences for murder and kidnapping are severe: Life imprisonment or death for first-degree murder. Ten years to life for second-degree murder. Kidnapping carries a twenty-year maximum term. The wholesale distribution of narcotics can get you ten years to life with a $4 million fine; trafficking in human organs, a five-year prison term and a $50,000 fine. In California, prostitution has a six-month jail term and a $500 fine, drug possession one year and $1,000.

So where does violating the Endangered Species Act rank? A one-year maximum prison term and a $50,000 maximum fine. A little more than for prostitution or possessing drugs (laws regularly broken by politicians, celebrities, and ordinary citizens), but nowhere near kidnapping or the distribution of drugs.

In 2008, a 46-year-old from Lucedale, Mississippi, was arrested for killing a Louisiana black bear, and then transporting it across state lines in violation of the Lacey Act. He pleaded guilty, served a 30-day sentence, was fined $5,000, and was ordered to pay $10,000 to the Bear Education and Restoration Group of Mississippi. In 2009, a 78-year-old Kauai man pleaded guilty to the fatal shooting of a pregnant Hawaiian monk seal, claiming he only wanted to scare it away; he received a 90-day jail term. Contraband wildlife, including elephant ivory and the pelts of tigers and jaguars, was found on the yacht of a Russian billionaire in Florida; his company paid $150,000 in fines. A New Mexico man, found guilty of unlawfully possessing a Mexican gray wolf, admitted shooting it but claimed he hadn't known it was a wolf. As part of a plea agreement, he was given six months in jail and had to forfeit his rifle.

The trouble with such penalties is that they're aimed at a mostly outdated threat: the intentional killing of a listed species. There have been few prosecutions for habitat alteration, though in 2010, a San Francisco man did plead guilty to violating the Endangered Species Act by closing off a dam on his property. His action prevented steelhead trout, listed as threatened since 1997, from returning to their spawning grounds.

You can't protect a species without protecting its habitat. And you can't protect habitat without conserving all of its species, including humans. As Andy Dobson, an ecologist at Princeton, said, "Perhaps all species are of equal value."

7

Here's how it was supposed to work: List a species. Come up with a plan to remove threats to its existence. Maybe stop the hunting or get rid of the pesticides. Protect it, preferably on public land. Let the species recover. Delist it. Move on.

If only.

Some species have indeed done very well under this type of protection. *Alligator mississipiensis,* a member of the Class of '67, was a textbook case. The soft and pliable belly skins first became fashionable in Europe during the mid-nineteenth century. In the next hundred years, more than ten million alligators were killed for the sake of purses, briefcases, belts, wallets, and boots. About fifty thousand hides were shipped annually from Florida alone. The newly hatched were killed, stuffed, and dressed up as tourists. Newlyweds on their honeymoon could buy a gator bride and groom. Nestling heads adorned key chains. Baby gators became popular pets, and then escaped into the sewers of New York City's imagination.

By the 1960s, only 5 percent of the original population was left in the Everglades, and at long last, in 1969, interstate trade of alligator leather was banned.[1] But the threat wasn't only from hunting. The draining of the swamplands had an immediate effect not just on the alligators but on the local Seminole Indians as well. As one elder related, "In the old times we could paddle our canoes for many days and hunt the deer and the alligator. Now the white man has drained the Glades with his canals to make the fields for his tomatoes and sugarcane. Our canoes cannot run on sand. . . . And the deer and the alligator each day go farther away."[2]

When the market for alligator hides, deer pelts, and the plumes of wading birds began to dry up, some Seminole hunters switched to Florida's perennial crop: vacationers. In the 1930s, the Tamiami Trail between Fort Myers and Miami had at least ten Indian "tourist villages" where visitors could see Seminoles weaving baskets and wrestling alligators. The latter was no tribal tradition. The first wrestler was Alligator Joe, a Florida entrepreneur born Warren Frazee who sported a long mustache and a wide-brimmed hat as he showed crowds on the beach how to fight "crocodiles" in the Atlantic surf. (The "defeated" reptile was tied up at the end of the show and hauled to the next location.) He opened his first roadside attraction in Palm Beach around 1895, followed by an alligator farm in Miami in 1911 that was the largest in Florida, the Disney World of the early twentieth century. By the time the Endangered Species Act was passed, it was rare to see a live alligator that was not on exhibit.

The alligator was a model for the Endangered Species Act. With hunting banned and habitat preserved, it had fully recovered by 1987, less than twenty years after it was declared endangered. Florida now hosts up to two million of them, Louisiana about 750,000. You see them by the roadside, in ponds, golf-course water hazards, and, on occasion, in backyard swimming pools. A twelve-footer was removed from a creek near the Veterans Affairs Medical Center in Miami, where it may have been feeding on urban wildlife—raccoons and possums, and the carcasses of sacrificed animals tossed in the water after Santería ceremonies.

Species can be delisted for three reasons: extinction, taxonomic reclassification, or recovery. The alligator's recovery has been widespread and long-lasting, but it's still listed. To the untrained eye, its skin is very similar to the American crocodile's and that of other endangered members of the order. Not everyone can discern the alligator's umbilical scars or find the telltale small dimple in crocodile skin. The alligator remains on the list to help monitor the sale of other crocodilians.

How did it come back so fast, while species such as the American crocodile still languish on the list? "Alligators are capable of becoming secretive if they're under pressure," Florida state herpetologist Paul Moler told me. They may have been lurking in inaccessible swamps, creeks, and bayous for decades. Females can be prolific, laying about 30 or 40 eggs per clutch; about half of these hatch and reach the water.[3] Once the persecution stopped, it helped that the alligator was loud and brash. The eighteenth-century naturalist William Bartram described its call as "very heavy distant thunder, not only shaking the air and waters, but causing the earth to tremble." A large chorus of alligators convinced him "that the whole globe" was "violently and dangerously agitated."[4] (Nowadays, a truck, loud music, or even the deep bellow of a tourist can set an alligator off.) Their terrifying roars may help distant pairs come together in the spring, reducing what biologists call the Allee effect: when populations get too small, it may be hard for an individual to find a mate.

Most predators in Florida are secretive—if you're lucky you *might* get to see the paw print of a panther in the panther refuge. But not alligators. Basking is an essential activity for these lie-and-wait reptiles. Stopping by an alligator pond, you might think of a landscape painting—alligators motionless, at rest among water hyacinths—but the crocodilians are actually hard at work. With 70 to 80 conical teeth, sensory organs around the mouth that function as motion detectors, and a bite of about three thousand pounds per square inch, they can eat just about anything that gets too close: shad, bowfin, gar, birds, or muskrats. They can crack red-bellied turtles like croutons. A full-grown one can grab a white-tailed deer on the bank and hold it underwater until it drowns. The real work is in the digestion—they need to maintain a body temperature of at least 81°F—at lower temperatures, prey ferments in the stomach and the immune systems break down. The higher the body temperature, the faster an alligator grows.

When I visited Fakahatchee Strand, a gently sloping limestone plain just northwest of the Everglades, it was 90 degrees in the shade—alligator growing season. Two adults slumped on the edge of a swamp pond. Beneath the boardwalk, three juvenile gators, each about a foot-and-a-half long, rested on a cypress plank. Another half a dozen hid behind a log.

A little blue heron, a kingfisher, and a snowy egret fished beside the gators in what looked to be the only watering hole around. A vulture perched on a dead cypress, lording over the scene. A red-shouldered hawk swooped, trying to steal a fish from the little blue. I fumbled for my binoculars and sent my knapsack—and laptop—tumbling into the mud. The juveniles dove beneath the log, uttering *yeonk! yeonk! yeonk!* as they fled. I hesitated for a moment—the laptop!—and jumped off the boardwalk. As I grabbed my bag, one of the two adults—maybe seven or eight feet long—eased into the water, its prominent brow rising from the dark.

I pushed the bag onto the boardwalk. Well, *it* was safe. I tried to get my belly up to the planks, my feet splashing the black water. The boards were slick from an afternoon downpour. She—for it must have been the mother—was only a few yards away, eyes glued to the log. I dug my fingers in between the planks and hoisted my chest onto the walk. *Don't think about deer.* I elbowed my way back up. Be still, my racing heart. The Endangered Species Act *was* working.

Several other species have followed the alligator model of recovery. The gray whale now numbers in the tens of thousands, enough for a resumption of traditional whaling along the West Coast to be considered. In 1999, a single whale was taken by the Makah Nation on the Olympic Peninsula. Although it is considered recovered and no longer listed under the ESA, the whale remains protected by the Marine Mammal Protection Act. Battles over the rights for the Makah to hunt gray whales continue.

Though the shortnose sturgeon, a member of the Class of '67, lives upstream of one of the largest cities in the world, it appears to have recovered in the Hudson River. Direct harvesting was suspended, dredging and pollution minimized in its range, and numbers have risen 400 percent. Fish tagged by biologists in the seventies were recaptured in the nineties. Down south on the Suwannee River, Gulf sturgeon are making a comeback—doing so well that some feel they now put people at risk: when the two-hundred-pounders leap out of the water, they can strike boaters or cause them to veer and crash. Like the alligator, these were perhaps the easy ones. As I would discover, many other species would require the work of a far more interventionist squad.

Back in Florida, the pond had settled down. The animals surrounding it depended on gator holes during the dry season. Fish survive in them—as long as they can avoid the gators and the wading birds. Raccoons, deer, and other creatures come to the banks to drink at their own risk. Alligators were wetland engineers long before humans arrived, expanding their deep black reservoirs year after year—which isn't to say you could see much of this engineering work going on during a typical day.

I watched the gators at Fakahatchee. A cloud passed overhead. A contrail evaporated from the sky. The juveniles called for their mother, who had only her watchful eyes above the water. Even the vulture on the snag above seemed in no particular rush for something to die.

The system of refuges that protects the alligator and hundreds of other listed species was started more than a hundred years ago. By the late nineteenth century, plume hunters had killed so many pelicans that there was only one nesting ground left on Florida's east coast, a five-acre mangrove island about 30 miles south of Cape Canaveral. Fortunately for the birds, Paul Kroegel, an immigrant from Germany, had built his home on an ancient shell midden overlooking the island. Word got out about how he sailed out to Pelican Island every day with a double-barreled shotgun and stood guard. In 1903 when Theodore Roosevelt heard the story, he asked his attorney general if there were any laws that enabled a president to establish a bird reservation. When Philander Knox said no, Roosevelt asked if there were any laws preventing him from declaring sanctuaries. His solicitor couldn't think of one. "Very well," Roosevelt responded, "then I so declare it!"[5] Kroegel was paid a dollar a month to manage the first national preserve to protect wildlife.

Roosevelt went on to create 53 more, along with 16 national monuments and 5 national parks; all told, he set aside 234 million acres of American wilderness. Many of these early refuges were purchased to protect herds of large mammals such as bison and elk—he was a big game hunter at heart. Lack of funding kept the program from growing after Roosevelt left office.

The money that would eventually be raised to purchase more than five million acres of wetlands got going with a mumble. Some say South Dakota Senator Peter Norbeck, who had a thick Scandinavian accent, had forgot to wear his false teeth when he gave a speech in favor of funding waterfowl refuges. Few in the Senate understood what he was saying, but many knew he was dying of cancer and his proposal passed unanimously.[6] The Migratory Bird Hunting Stamp Act, or the Duck Stamp Act, would bring in more than $650 million in revenue.

Even after the money started coming in, there was little guidance on how to spend it. To oversee the purchase of waterfowl habitat, Franklin Roosevelt gathered an advisory committee that included Aldo Leopold and Jay Norwood "Ding" Darling. Darling—the head of the Biological Survey, a political cartoonist, and "the best friend a duck ever had"—designed the first Duck Stamp, a pair of mallards landing in a pond. Back in 1934, it cost a dollar; now it's fifteen. Soon after the committee formed, Darling persuaded John Clark Salyer II to stop work on his PhD at the University of Michigan and start looking for suitable real estate. Salyer had a fear of flying, but the days of surveying by rail were long gone, so he crisscrossed the country by car. Within six weeks of being hired, he had mapped out plans for six hundred thousand acres of new refuge lands—and put eighteen thousand miles on his government-issued Oldsmobile. Darling admired Salyer's ability to make complex water analyses on the trunk of the battered old car. The days were for working, the nights for driving. Salyer told one writer that he saw Mount Shasta in the moonlight four times before he saw it in the light of day.[7]

Salyer helped protect wetlands in California's Central Valley, the mountains of Montana, the North Dakota prairies, Maine's North Woods, and along the coast of Delaware. He tirelessly visited the refuges, insisting on improvements along the way. One of his bosses said, "We could trace his progress cross the country by the anguished wails of the regional supervisors." During Salyer's tenure as head of wildlife refuges, he increased the protected acreage from less than 2 million to almost 30 million acres.[8]

The trouble with the refuge system, especially after Salyer retired in 1961, was that it lacked an overriding vision. The 1966 National Wildlife Refuge Administration Act allowed individual managers and the Secretary of the Interior to permit "any use of any area within the System for any purpose" as long as it was compatible with the original goals of the refuge. Some reserves allowed oil and gas extraction, others bombing practice; many had mines or grazing livestock or an active timber industry. Since the passage of the Endangered Species Act, more than 60 refuges have been created expressly to protect listed species. By uniting the system and bringing in a professional staff, the Act helped the Fish and Wildlife Service refocus on wildlife.

In 1997, Congress passed the Organic Act, which helped resolve some of the remaining conflicts. The Secretary of the Interior was required to ensure and maintain the biological integrity, diversity, and environmental health of the refuge system. Conservation plans were required for all refuges every fifteen years. Mining, lumbering, and grazing were to take a backseat to wildlife. Although refuges were rarely at center stage in environmental battles, the old phrase "any purpose" remained a threat. Republicans in Congress tried to open up the Arctic Wildlife Refuge to drilling throughout the Bush administration but they never found the votes.

In contrast to national parks, which attract visitors who tend to be educated, affluent, and predominantly white, a wider range of people can be found at wildlife refuges.[9] About a third of the refuges offer hunting, and more than half provide fishing—attracting people who may have less education and lower incomes. The celebrity parks of Yellowstone and Yosemite may take some getting to, but anyone can take the subway to Jamaica Bay Wildlife Refuge in New York City. Inner-city residents watch birds, catch fish, some poach turtles or clams, a few even shoot there. While working for the New York City Audubon Society in the 1990s, I was in charge of a group of City Corps volunteers (a precursor to AmeriCorps) who helped restore and survey a small park bordering the refuge. On our first day at the site, which was a small sanctuary across from a landfill, we heard a woman scream. Then someone yelled, "Stop!" Two men ran by with guns. I yelled at the volunteers—city kids in their teens—to get down. And then we saw the trailing edge: a soundman and a camerawoman. Someone yelled, "Cut!"

Refuges are finally being recognized for their contribution to a healthy economy. The Department of the Interior estimates that National Wildlife Refuges brought in about $1.7 billion in 2006, supporting almost twenty-seven thousand jobs in the private sector.[10] Habitat protected for endangered species generates at least $180 million a year from tourism, along with tens of millions of dollars of employment income, tax revenue, and private-sector jobs.[11] These are the direct benefits, in the form of tourism, but refuges also provide substantial nonmarket ecosystem services. Many coastal refuges, set aside for wintering waterfowl and birds migrating from the tropics, have marshes that act like sponges, protecting cities by absorbing water during storm surges. Prescribed burns on forestlands reduce the risk of a catastrophic blaze. After Florida swamps were drained in the 1960s, refuges at Big Cypress and Fakahatchee were set aside in part to prevent the intrusion of salt water into the drinking supply of hundreds of thousands of Gulf Coast residents. It was a classic win-win situation: reliable fresh water for residents, more habitat for the region's swamp things. The lands and the services are protected by the government: the benefits accrue to the public and are not at risk to private interests that can profit by selling off the last tree or fish. It's part of our natural infrastructure, our natural capital.

Although it's a boon to local economies—and vital to the existence of many species—the refuge system isn't big enough to maintain the ecological processes needed for the recovery of most endangered species, nor will it prevent all extinctions. Only about a third of the more than five hundred listed animal species are found in national wildlife refuges.[12] With the exception of a few wild game species, long-term conservation will require viable populations on public *and* private land. And sometimes even that's not enough.

8

On June 30, 1954, a solar eclipse darkened the skies in northern Canada. A fire patrol in the air that day recorded that it spotted on the boreal plains of the Northwest Territories a pair of large white birds with a smaller tawny one. The sighting was unusual for that part of Canada, and word got out to ornithologists.

Robert Allen, the National Audubon Society's first director of research, had spent more than two years in the field, hoping to make just such an observation. In 1945, he had been assigned to gather sufficient scientific knowledge to save the highly endangered whooping crane. He had observed them on their wintering grounds in Texas, but their summer breeding grounds remained elusive, even after he had completed extensive surveys over Alaska and western Canada.[1] With the help of the fire patrol, in 1955, Allen finally discovered several whooping crane nests in Wood Buffalo National Park. The existence of the species was precarious—it had one wintering ground, and this was its only known breeding area—but Allen was relieved that the birds had chosen a remote and protected site in which to raise their offspring: the largest national park in Canada, established in 1922 to protect free-roaming bison.

The birds had almost completely disappeared. After decades of market hunters killing them for the meat and long white feathers and of egg collectors robbing their nests, the entire population of cranes had gone from 15,000 or 20,000 in the early nineteenth century to 1,500 in 1870—and to 22 in 1941. Much of their marsh and meadow habitat in the Midwest had been lost to agriculture.

Only sixteen cranes wintered in the protected coastal wetlands of Texas. Salyer had seen the value of the marshes for the cranes and other waterfowl, and established the Aransas National Wildlife Refuge there in 1937. A small nonmigratory flock in southwest Louisiana had almost been wiped out by a hurricane that hit the state in 1940. The half dozen birds that remained did not reproduce; by 1948 a single bird was left. Ornithologist David Ellis told me that the biologists at the time figured, "Well, why just let him die?" They flew two helicopters into White Lake, about 125 miles west of New Orleans, in March 1950 and captured Mac, as he came to be called. They released him at Aransas, where he was attacked by the local cranes; later that summer, his carcass was found on the refuge. The last of the nonmigratory birds was gone.

The Fish and Wildlife Service began holding press conferences to announce the annual whooper count: 28 cranes in 1955, 24 in 1956, 26 in 1957. The public was captivated by this vanishing species: a leggy white skyscraper of a bird, four-and-a-half feet tall with a bright red crown atop a black facemask. People followed the numbers like a sporting event: the birds were in danger of a permanent shutout. Although whooper numbers rose during the following ten years, only one population remained; a single catastrophe could wipe out the species.

In the late 1950s, a group that would become the Whooping Crane Conservation Association coalesced around the idea that captive breeding was essential to saving the bird.[2] It had its critics. Robert Allen, for one, concerned that eventually all the birds would be taken from the wild, wrote, "It is not our wish to see this noble species preserved behind wire, a faded, flightless, unhappy imitation of his wild, free-flying brethren."[3] So a compromise was struck. Ray Erickson, a biologist at Fish and Wildlife Service, showed that although many cranes laid two eggs in a clutch, often only one survived. The removal of a single egg, he contended, would present little risk to the wild population.

So in 1967, as the initial list of endangered species was being drawn up, biologists at Wood Buffalo started the experiment. Ernie

Kuyt of the Canadian Wildlife Service found a pair of whooping crane eggs in a shallow pond surrounded by a sedge meadow. He had forgotten his collecting container and gently pushed the egg into a wool sock. It was the humble start of an ongoing experiment that would challenge our understanding of what it means to be a crane.

It was clear, perhaps long before the Endangered Species Act was enacted, that the alligator model wasn't going to work for every species. For some, habitat loss was the problem; others, like the whooping crane, had been reduced to such low levels that only a single vulnerable population remained. Individuals could suffer from inbreeding or retain behaviors that worked for large populations, but were devastating to small ones.

Most of the mottled pale-blue eggs removed from Wood Buffalo were relocated to Patuxent Wildlife Research Center in Maryland. Patuxent, the research refuge that Ding Darling and his colleagues helped establish in 1936, has been at the center of several groundbreaking environmental studies. In the 1940s, dichloro-diphenyl-trichloroethane, DDT, was seen as the answer to the long war on insect pests: dusted on clothes, it protected against lice and fleas and the diseases they carried; sprayed on ponds and marshes, it controlled malaria; spread by planes over fields and forests, it wiped out potato beetles, boll weevils, and other pests. But how did it affect the rest of nature? Researchers at Patuxent used Navy airplanes to spray a 117-acre tract of the refuge with two pounds of DDT per acre. The earliest studies found little effect on bird populations and egg development, but as studies began to accumulate, there was an increasing sense that DDT impacted nestling survival. Rachel Carson heard about the work in 1945 and pitched a story idea on it to *Reader's Digest,* telling the editor that the research was being done "practically at my backdoor."[4] But the magazine turned her down and she dropped the subject for more than a decade.

In the 1960s, the research station became the center for federal efforts to propagate and study endangered species, including whooping cranes and black-footed ferrets. A few years ago, I visited Patuxent in early March. The temperature was in the forties, the skies gray, the grass the color of an old kitchen broom. Kathy O'Malley, a twenty-year veteran of the whooping crane project, took me on a tour of the grounds. We passed the impoundments where acid rain studies were conducted in the 1980s, the fences rusting along the roadsides. We took a left at the corner of Whooper Drive and Sandhill Loop. A sign warned, "Crane Breeding Area: No Admittance."

As we approached the crane enclosures—large pens of about a quarter-acre each—we were greeted by a single bright bleat through the eight-foot-high chain-link fence. The whooper's trachea, coiled in its sternum like a French horn, gives the bird's guard call (a classic *whoop*) a deep resonance; a sandhill crane in a nearby pen echoed the call in a raspy voice, almost like that of a smoker.

There were 74 Florida sandhill cranes, 67 greater sandhills, and 54 adult whoopers when I visited. The sandhills, which now number about 650,000 in North America, have played a critical role in the restoration of their rare cousins. Soon after the whoopers laid their eggs, aviculturists removed the viable ones and handed them over to a pair of sandhill cranes, who incubated them for the first twenty days. These surrogates didn't actually raise the chicks—O'Malley and her colleagues didn't want the whoopers to grow up thinking they were mutant sandhills. In the later stages, most eggs were placed in old wooden incubators in a brick "propagation" building. Every few years, the successful surrogates were allowed to raise a chick of their own "to keep their motivation up," O'Malley explained.

At the first pen, O'Malley trilled to the ten-month-old chicks. The long-legged, rather drab birds responded with peeps and a purr. Nearby, a fifteen-year-old crane named Lonely strutted across his pen, slowly displaying his red crest: bare skin with tiny hair feathers above a black forehead, yellow eyes, and a foot-long beak—a formidable weapon. Lonely stretched his wings, revealing a dark cloud of black primary feathers. After this slow-motion display of aggression, he pointed his bill at the sky and bugled. Another male practiced a mating dance, hopping on spindly legs almost three feet long. He was slightly off-kilter—one wing was kept clipped until a potential mate was introduced. This was crane life in captivity—and practice for parents, foster parents, and even for the birds that would eventually be released. O'Malley told me that Lonely was an excellent role model, imprinting on youngsters what it was like to be a whooper.

Even before they hatched, the eggs were exposed to the sounds of adult cranes to accustom the chicks to their own kind. Handlers kept a vow of silence around the eggs and young birds so they didn't become used to human voices. But the cranes did have to get used to one man-made sound: engine noise.

For most of aviation history, airplanes flew too fast for birds. Pilots might smash into them—the National Wildlife Strike Database records 38,961 aircraft collisions with birds between 1990 and 2004—or gaze at them from afar during takeoff or landing, but birds (especially the larger ones that flew the thermals) moved too slowly for any constructive interaction.

In the early 1990s, several hang-gliding pilots realized that ultralight planes could fly at a bird's pace. One, Jeff McNeely, made an award-winning film with his trained red-tailed hawk. In 1993, Bill Lishman, an artist, and Joe Duff, a photographer, trained several Canada geese to imprint on their ultralights. They then led eighteen goslings on a 680-kilometer journey from Purple Hill, Ontario, to Virginia. As the pilots hoped, the birds learned the migration route along the way. They all survived the winter; thirteen flew back to Purple Hill the following summer. After this success, the pilots trained trumpeter swans and sandhill cranes. Cranes led by ultralight were attacked by golden eagles; others led by truck—actually a US Army ambulance—became tame: one ended up in the front yard of a Mrs. Crane in Idaho; another moved in with a domestic ostrich.[5] The trainers refined their techniques to make the birds warier of humans.

They flew at higher altitudes to avoid eagle attacks. By the end of the decade, the team was ready. It approached the whooping crane recovery team with a seemingly crazy idea: to lead the world's most endangered crane across the United States in a tiny aircraft, and leave them to find their own way back.

Ornithologists once felt compelled to shoot, stuff, and eat their subjects in order to understand them; latter-day crane biologists learned to embody their favored birds. David Ellis, a research behaviorist at Patuxent, began to prep the cranes for the journey; dressed in long white shrouds to mask their human forms, aviculturists used crane puppets to feed and train the young and to prevent the birds from associating people with food. Think *Star Wars* meets Casper the Friendly Ghost. This imaginative leap on the part of the biologists— and perhaps on the part of the cranes themselves—led to the establishment of a new migration corridor east of the Mississippi. By keeping this flock separate from the one in the Midwest, a new route would limit the opportunity for disease transmission between the two groups and allow each flock to develop its own traditions and, eventually, its own genotypes.

Why fly Wisconsin to Florida? Refuges, of course. At the northern end of the migration route, Wisconsin's Necedah National Wildlife Refuge was chosen because it is in the midst of more than two hundred thousand acres of marsh—away from hunters, power lines, and pesticides. Salyer had purchased the land during a cross-country jaunt in 1939; it was a good deal—the area was under eighteen inches of water for much of the year, too wet for farming but perfect for wading cranes and waterfowl. Whoopers like to retreat to these deeper waters at dusk in an attempt to avoid bobcats and other crepuscular predators.

On the southern end, Florida lobbied for the whoopers. Sandhill cranes were already a welcome sign of winter. "When they arrive," a bookseller in the state told me, "it's the happiest sound." An even taller husky-voiced bird, coaxed in by ultralight, would add to the peninsula's winter chorus—but crane proponents had more than calls and conservation in mind. In 2006, tourists spent over $3 billion in the state to see birds, alligators, manatees, and other wildlife. They supported fifty-one thousand jobs (about as many as Walt Disney World) and $1.6 billion dollars in salary and wages, a figure comparable to all the money spent on golf equipment across the country. Many visitors used part of a two-thousand-mile birding trail with its more than 445 stops.[6]

First, the birds had to become established and given a little solitude. Chassahowitzka National Wildlife Refuge on the Gulf Coast—a remote location with abundant blue crabs, a favorite meal for the wild whoopers in Texas—seemed to be the perfect place. (Salyer? Yep, he had bought it in 1941, though by then he was moving up the ranks of the Service.)

At 7:15 a.m. on October 17, 2001, two ultralight planes and eight cranes lifted off through the morning mist of Necedah. Within minutes, one of the cranes left the group and landed several miles away. Biologists with handheld radios spent the day looking for it, eventually getting a signal from a densely wooded lakeshore. They donned their crane costumes, played a recording of some contact calls, and drew in the errant bird. It was hooded and driven to rejoin the flock. High winds would ground the group or force it to fly low. The birds shied away from the noise of the interstates, circling back to wherever they'd been penned overnight. One crane died in a windstorm. Two of the birds that made it to Chassahowitzka were killed by bobcats. But five birds survived and returned to Necedah on their own the following year.

Every October since, a caravan of birds and people takes off from Wisconsin on a journey that is part biological expedition, part traveling sideshow. A shrouded pilot, playing taped calls from his ultralight, leads a line of cranes. RVs stocked with costumes, puppets, temporary pens, and food pellets follow on the ground, with crews leapfrogging from site to site, setting up overnight pens on farms and ranches. In decent conditions, the flock travels about 35 miles an hour; with a tailwind, they can cover two hundred miles in a day.

Lara Fondow—a graduate student and research associate with the International Crane Foundation in Baraboo, Wisconsin—took me out to see the Florida cranes. She pulled up to my hotel in Homosassa in an enormous ghost-white Suburban, a through-the-roof Yagi antenna poking up from the cab like a giant fish skeleton. As we headed out to the cranes' last known location, a three-thousand-acre cattle ranch in Pasco County, Fondow multitasked. She scanned for radio signals, holding a vice grip attached to the antenna, and talked to her boss via cell phone, the truck left to negotiate itself down the quiet Florida road.

As we turned onto the ranch, the telemetry receiver began to emit a soft but steady flow of chicklike chirps. Fondow quickly located a pair of whoopers on a pasture just northwest of the dirt road. "It's got a snake!" One of the tall birds shook its red-capped head, snapping the reptile like a whip, before it wolfed it down. In marshes, adult birds filled up on blue crabs, but out here in the pastures, they consumed voles, mice, grasshoppers, acorns, and the occasional snake.

These were closely monitored, human-raised birds. Like the red-cockaded woodpeckers, the cranes wore variously colored bands for quick ID. Radio transmitters allowed Fondow and her colleagues to track them from these Florida winter grounds to Necedah, unless the birds went out of range. In 2005, a lone female wound up in a field near Lake Champlain, not far from my home in Vermont. That same bird returned in May 2006, this time followed by a younger, inexperienced bird. A group of seven Florida whoopers traveled to South Carolina in 2005, feeding in salt marshes there for much of the winter. The experimental run from Necedah to Chassahowitzka was a nice start, but apparently a few birds had plans of their own. If the experiment works, who knows where they'll all end up?

Fondow and I walked through saw palmetto to a small group of cranes at the far end of a pasture. They appeared to be thriving amidst the barbed wire and beef cattle. "All of our birds are on cattle ranches," Fondow said. "They're the only open lands left."

Whoopers are formidable, aggressive birds, easily capable of fighting off domestic dogs, but they have a problem with cats—wild cats, anyway. Bobcats have been the primary predators of introduced whoopers. Grazed land doesn't provide much cover—good for cranes, not so good for bobcats, which hunt by stealth, sneaking up on their prey in the shadows thrown at dusk and dawn. Cattle may be considered a scourge in the arid West, but these rangelands could be the whooping crane's best hope against the southern sprawl of housing developments, assisted-care facilities, trailer parks, and strip malls. Eat Florida beef, save the cranes?

During the trip to Florida, I took state biologist Steve Nesbitt's advice and drove out to Lake Kissimmee, just south of Orlando. At the time, the stats had 87 resident whoopers spending the entire year on the open palmetto prairie around the lake. Six pairs had hatched chicks, four of which fledged. It's a sign of the crane's continued precariousness that we're still keeping tabs on each nest, each bird. Much of the population survives in captivity, with occasional releases. In the 1970s and 1980s, biologists tried to introduce a flock to the Rocky Mountains. The Wood Buffalo whoopers, fostered by sandhills, crashed into power lines and died of disease.

After passing through cattle ranches and palmetto scrub, I reached the vast shallow lake at about 5:30 p.m. It had taken on the color of burnished copper—or the subtle glow of an agar plate, the cultured medium of millions of lab experiments. As I lingered by the shore, a black bull snorted, backing me closer to my rental car, despite the barbed wire that stretched between us. A bird I couldn't quite ID, a silhouette with a large bill, perched on a fencepost. Five black skimmers busied themselves slicing the sun to tatters. The crescent moon over Kissimmee flashed a Cheshire grin.

The silhouette lifted its wings and circled back to the live oaks. It raised its white head from the shadows—a bald eagle, the embodiment of conservation success. When they joined the Class of '67, eagles were so rare that the Mason Neck National Wildlife Refuge in Virginia was established for the sake of a single bird. Habitat protection and the banning of DDT helped, as did a captive-breeding program that released eagles throughout the country. There are now more breeding pairs along the Chesapeake than there were in the entire lower 48 when the Endangered Species Act was passed. There are breeding eagles in every state on the continent. The bald eagle was delisted in 2007.

The eagle landed with its mate, but it was getting dark, and I began to think about finding a hotel. Then I heard a distant call.

Three birds, long necks outstretched, approached from a nearby pasture. I wasn't sure if they were sandhills or whoopers until I saw that surprising hump of a back, rising above the tree line, and heard another call—part bugle, part hyena—that heralded the arrival of *Grus americana*. Gliding west, the whoopers appeared powder blue against the copper sky, their black legs dangling like roots. I watched the trio and a dozen sandhills fly to the safety of the marshes just before sunset: bobcat breakfast time.

I thought of what O'Malley told me after the tour of Patuxent. "The first time I saw them taking wing in the wild was one of the most special moments of my life." Her eyes had welled up. All of these whoopers, like the others I saw in Florida, were raised by O'Malley and her colleagues. But after 30 years and the release of more than 250 birds, no new breeding populations have been established—not in Necedah, in Florida, or anywhere else in the United States. And despite all the birds' training, a catastrophe can strike anytime, anywhere. In February 2007, seventeen birds of the 2006 migratory class were killed in a storm, probably drowned in a tidal surge. Only one chick survived; found with a pair of sandhills foraging far from the pen, it died several months later. In addition to the storm, bobcats, alligators, coyotes, eagles, power lines, and aircraft took out more. Still, the whoopers had come a long way from the sixteen Texas birds of 1941. That flock—the only natural one left in the world—now numbered 268.

The recovery team has stopped supplementing the Florida non-migratory flock. Despite the release of 289 captive-raised whooping cranes, 68 nesting attempts, 31 hatched chicks, and 9 fledges, the population was suffering from high mortality and low reproduction. There were just 29 birds in 2010. All of the year's chicks had died before they reached a month old. It looked as if this experiment had failed. As Nesbitt told me, they just couldn't justify spending so much time and money to feed bobcats.

Endangered species recovery often requires risky, contentious, and imaginative strategies. Whooping cranes were relatively easy: excess eggs could be removed from a single well-protected flock without too much risk to the adults. The condors of California were another matter. In 1986, one of the last remaining female condors of breeding age died of lead poisoning, despite heroic efforts to revive her.

It was a pivotal year for conservation. In April, Michael Soulé organized the Society of Conservation Biology and founded its journal, launching the discipline. The term *biodiversity,* which didn't exist when the Endangered Species Act was passed, was coined as shorthand for "biological diversity" in internal plans for a forum held at the Smithsonian Institute in September. Most practitioners quickly embraced the goal of protecting biodiversity, which was as all encompassing, spanning from genes to ecosystems, as it was difficult to define. Within a few years, new journals sprouted up with the term: *Canadian Biodiversity* (1991), which changed its name to *Global Biodiversity* two years later, *Tropical Biodiversity* (1992), and *Biodiversity Letters* (1993).[7]

The emergence of conservation biology, with scientists being trained in the tools of protection and restoration, helped support the decision to capture all of the California condors left in the wild in 1987. It was dangerous; what if it failed and the last of the birds died in captivity? It was also bitterly controversial. The famed conservationist David Brower called condors "soaring manifestations of the place that built them and coded their genes." Condors were "only 5% bones and blood and feathers . . . the rest is habitat." Condors in zoos, he said, were like "feathered pigs."[8] The National Audubon Society filed a lawsuit to prevent the roundup of the last wild birds; at least a few condors, they argued, should be left in the wild to serve as vanguards for a future captive-bred population.[9]

There had already been another roundup on the other side of the continent. In 1979 and 1980, the last six dusky seaside sparrows were collected from St. Johns National Wildlife Refuge in Florida, former ranchland that had been acquired to protect the birds, whose native marshes had been drained, burned for pasture, or destroyed by highways. All of the birds captured were male. With no pure females left, Herbert Kale, the Florida ornithologist who had captured them, proposed mating the males with hybrid dusky and Scott's seaside sparrows, a related bird found in northwestern Florida. The Fish and Wildlife Service balked; the Interior Department's solicitor had handed down a series of legal opinions that the Endangered Species Act covered only pure species. After two years, the birds were taken to Walt Disney World's Discovery Island, which offered to pick up the tab for the breeding program. Some birds were incompatible; by then, most of the males were too old. They built nests and laid eggs, but none survived. Several birds were killed in storms, others were eaten by rats, and a few just disappeared.[10] The last pure male, known as Orange Band, died on June 16, 1987.

The sparrow was in the Class of '67, yet it never reached the priority level of charismatic species like the condor. (It didn't help that it was downgraded to a subspecies in 1973.) The cordgrass marshes around Merritt Island, which had been poisoned by DDT and drained for mosquito control, have since been restored, but there's a new silence. The sparrow's raspy, three-syllable whir, a haunting "Look at me," is gone. Full stop.

The experience with the dusky was hardly an endorsement for another complete roundup, but it was clear that in the sparrow's case, the managers had waited until it was too late. The condor biologists didn't want to make the same mistake. Once the last bird was plucked from the wild, on Easter Sunday 1987, public sentiment began to change, and even former critics supported the program.[11]

Beginning in 1992, condors were gradually reintroduced; by 2000, the editors of *Audubon* acknowledged the need for intervention to save the species: "Only by interacting with nature can we come to appreciate it, understand it, and, we hope, preserve it. What we have learned from the condor may help us to alter our behavior enough to save these birds.... At the very least, it may prevent us from killing off another symbol of the wild."[12] The first wild nestling fledged in 2003. By 2010, there were more condors in their native habitat than there were in captivity (just barely, 192 versus 189). It's a precarious success—the condor is likely to rely on conservation efforts for a long time.

At the same time as the condor and sparrow, another symbol was taken from the wild. First described by Audubon and Bachman in 1851, the black-footed ferret had once been a wildly successful species, feasting on the thousands, perhaps millions, of prairie dog towns that populated the Great Plains from Mexico to Canada. When cattle were moved into the West, ranchers and their allies in the federal government eradicated as many prairie dogs as they could with arsenic. Then sylvatic plague—the Old World's Black Death, lethal to humans and to many mammals—arrived in San Francisco in the early 1900s and spread throughout the West, devastating prairie dogs and their predators. The black-footed ferret was a member of the Class of '67, but little was done to protect the last known population in South Dakota, which blinked out in 1974, a year after the Endangered Species Act was signed. Many believed the ferret was extinct.

Then, in 1981, John Hogg, a Wyoming rancher, heard his dog barking in the middle of the night. The next morning he found an animal with a broken back in his yard. It looked like a mink with a black mask and feet. After he showed it to his wife, Hogg threw the old scrap that even the dog wouldn't eat over the fence. Later, she thought the strange animal might deserve a place on the mantel, so John climbed over the fence and brought the thing to town. The taxidermist looked at the punctured flesh and broken bones and said, "My God, it's a ferret. I can't touch it. They're an endangered species."[13] Hogg's dog had retrieved the black-footed ferret from extinction.

The animal was traced to a neighbor's property, the 120,000-acre Pitchfork Ranch, just east of Yellowstone. It was classic big sage country, Marlboro country—several cigarette ads had been shot here. The area soon became the center of an intense rescue effort so popular with government biologists and reporters that Lucy Hogg put Ferret Nuggets (tasted just like chicken—because they were) on the menu of her café. The owner of the ranch, Jack Turnell, might have been shocked by the attention at first, but he turned out to be an excellent steward of the land.

For a few years, the animals did well, reaching a population of 120 individuals, before plague hit the prairie dogs and canine distemper ravaged the ferrets, bringing them perilously close to extinction again. "There were some people that wanted them to go out with dignity," Mike Lockhart, who was with the Fish and Wildlife Service, told me. "Somebody finally had to put their foot down and say, 'No, we're going to bring ferrets into captivity.' When they did that, it was just about too late." The Wyoming Game and Fish Department, initially opposed to removing the ferrets from the wild, took the lead on captive breeding and field recovery; by most accounts it wasn't up to the task. One biologist called the recovery effort "a bureaucratic misadventure . . . in uncharted waters."[14] (Others weren't so kind.) All of the first six ferrets captured succumbed to distemper and died. Things continued to move at a snail's pace, one of the biologists involved in the project told me. By now, there were only eighteen individuals left in the world.

One of these, Scarface, was captured in 1987. He must have been a dominant male for he bred with the females right away. Thirty-eight kittens were born in 1988, 58 in 1989, and 66 in 1990. And things improved at the agencies, too: A recovery implementation team was established in 1996, forming a network of federal and state agencies, Indian tribes, and NGOs. Committees appointed by the team provided management direction, partners submitted proposals and comments on the reintroduction of species, and interagency rancor was reduced.

Ferrets were first reintroduced in Wyoming in 1991 and in the Badlands of South Dakota in 1994. The first kits were born in the wild in 1992; their populations have been increasing ever since. But things haven't always gone smoothly. Lockhart left the service in 2008, "falling on his sword" as he described it, to protest the Fish and Wildlife Service's unwillingness to support habitat conservation and the US Department of Agriculture's plan to poison prairie dogs in the South Dakota grasslands of the Conata Basin, home to the largest population of wild ferrets.

By 2009, black-footed ferrets had been reintroduced to South Dakota, Montana, Wyoming, Utah, Colorado, Kansas, New Mexico, Arizona, and Chihuahua, Mexico. There are now more than a thousand in the wild. "But some of those populations have never done well," Lockhart said. "Where we have really good populations are South Dakota and Wyoming, but both of those areas are still under threats of plague." Plague has three basic forms all caused by the same bacterium, *Yersinia pestis*: bubonic, which is transmitted by fleas and gradually takes over the lymph system; septicemic, when the bacterium invades the bloodstream; and pneumonic, through the lungs. Ferrets are at risk to all three—and one form can progress into another; exposure to the plague almost always kills ferrets within a few days.[15] Vaccinations have been developed to fight the disease and are given to the ferrets before they are reintroduced to the wild. The key will be vaccinating the prairie dogs themselves by developing bait that managers can use to inoculate entire colonies.

"Don't ever let a species get to the point that captive breeding is necessary," Lockhart warned. "That's fraught with all kinds of peril. . . . Our regional director [in the Bush administration] thought that as long as we have ferrets in captivity and we're cranking them out like a put-and-take fishery, it was okay." If you see a decline coming, he said, "Don't wait." Take the surplus while there's still a reasonable population size. And ultimately, you want to use wild populations as sources for reestablishment and gene flow, not captive ones.

Captive breeding is difficult, expensive, and usually only necessary once the species has all but disappeared. About five thousand terrestrial vertebrates, many of them amphibians, will probably need it to avoid extinction. Zoos currently have the capacity for about five hundred.[16]

Besides the shortage of space, evolutionary pressure is a major concern—inbreeding and rapid evolution can devastate captive populations. The longer animals stay in cages or tanks, the greater the selection intensity for these populations to adapt to a life behind bars or glass. When captive-reared steelhead trout, a threatened species, were released into native rivers, their fitness, or their reproductive output, was about 40 percent lower than that of their wild relatives.[17] The continued release of captive-reared fish to supplement a declining stock could lower the fitness of the entire population. How? Deleterious genetic alleles that were rare in the wild—partially recessive and masked from direct selection pressure—can become common in the safety of the hatchery. The increased frequency of these harmful genotypes, along with selective sweeps that reduce the diversity of nearby neutral regions, form the main basis of observed genetic adaptation to captivity.

Australian geneticist Richard Frankham notes that the most effective way to avoid this adaptation is to minimize the number of generations spent in captivity. Unfortunately, that's just not a practical option for many animal species. The Socorro dove, once native to an island off Mexico, has been extinct in the wild since 1972. The Wyoming toad hasn't bred beyond the confines of aquariums since 1991. The last wild Hawaiian crows died in 2002; only 78 remain in captivity. Single captive populations will quickly lose fitness and are at great risk of disease. But keeping them in several smaller populations, with gene flow occurring between captive and wild populations (if there are any wild ones left) can reduce inbreeding and evolutionary adaptation to life behind bars.[18] Whether species in captivity will ever have a place to return to—an ecosystem where they'll feel at home or at least survive—is another challenge, rarely under control of managers. As far as the ecosystem and the community are concerned, a species in captivity is already extinct.

9

The Florida panther was in a desperate state. Its habitat was shrinking, and highway deaths were commonplace. After a recovery plan was drawn up in 1981, the Florida Fish and Wildlife Conservation Commission began to capture cats for radio telemetry. They found fewer than 25. Of these, only about a dozen lived in protected areas; the rest were on private lands, mostly cattle ranches on the northern edge of what was left of the big cat's range. The remaining panthers looked closely related, sporting the kinked tails and cowlicks considered signs of inbreeding. More disturbing, more than 80 percent of the males were cryptorchid, with only one descended testicle or none at all. They had poor sperm quality and low testosterone levels. Many panthers had heart defects and high loads of parasites and infectious pathogens. When the author Charles Bergman visited South Florida in 1986, he wrote that the panther was a "deeply endangered animal, perhaps unsavable."[1]

For years, biologists claimed that the Florida panther was a unique subspecies of puma and should be preserved in its pure form, kinked tails and all. The panther appeared on license plates; school kids elected it the state mammal. But with only about two dozen left in the world, state and federal managers seemed faced with a choice: preserve the Florida panther as a unique "subspecies"—the genetic priority—and possibly watch it disappear, or introduce "non-Florida panthers" for "genetic enrichment purposes," protecting the felid's place in the ecosystem, if not the original cat itself.[2] Experts met at the National Zoo in Washington, DC, and at White Oak Plantation in Florida. Their models predicted a 95 percent likelihood of extinction within twenty years. Genetic rescues had worked for prairie chickens in Illinois, they argued, and for adders in northern Europe. Perhaps the import of new genes from the nearest population in Texas could rescue the panther as well.

But would hybrid cats still be protected by the Endangered Species Act? Historically, there had been gene flow between Texas cougars and Florida panthers, so how much would it hurt to reestablish that corridor, even if it involved the transport of a few cats from Big Bend? When the Fish and Wildlife Service determined that all of the offspring between Texas and Florida cats would qualify for protection, the stage was set. As with the ferret and the condor, the decision was controversial. Some argued that the kinked tails and isolation were part of being a panther in the twentieth century. Others claimed that only assertive habitat protection would allow the cat to increase its numbers and expand its range.

In 1995, Roy McBride, a big cat tracker from Alpine, Texas, captured eight female cougars in the Big Bend region along the Rio Grande. The cats were held in quarantine, checked for disease, and then flown to Florida on commercial airlines in crates built by McBride. Six of the animals were released among the core population just east of Naples, and two were let out in the Everglades. The results have been convincing: In 1990, 88 percent of the Florida population had kinked tails. By 2000, not a single first- or second-generation kitten born to the Texas females had the characteristic 90-degree crook in the last five vertebrae. The hybrid kittens had a higher survival rate than the purebred panthers, the females lived longer, and most bred. The population increased from about 30 cats to at least 95, and they expanded their range, entering habitats once thought unsuitable for panthers. Most of the young males had healthy reproductive systems—and the field biologists' job got tougher. According to Steve O'Brien, head of the Laboratory of Genomic Diversity in Washington, DC, the physical stamina of the new offspring was "better than that of their Florida panther parents and grandparents." A healthy cat could be more aggressive and tougher to tree.[3]

It doesn't take many individuals to reverse the trend toward inbreeding. By the 1970s, the gray wolf was considered extinct in Norway and Sweden. Hundreds of wolves had been killed every year on the Scandinavian Peninsula up until the nineteenth century, and bounties continued to be paid into the 1960s, when the canids were protected in Sweden, then Norway. It was almost too late; no young wolves were seen for years. But in 1978, the first successful reproduction in the wild in more than a decade was recorded in the peninsula. For the next twelve years, snow trackers followed the canids along the border of Sweden and Norway. Approximately nine hundred kilometers from the nearest known population in Russia, the pack (founded by just two wolves) reproduced every year but one, yet it remained in a single territory and never exceeded ten individuals, all close relatives. Wolves naturally try to avoid inbreeding, but if these didn't mate with relatives, they wouldn't reproduce at all.

In 1991, the population exploded. Ten packs were formed in fewer than ten years. The reason? The arrival of a single male, traveling on his own from the north, was enough to spark the recovery of the population.[4] The tiny range of the single pack has expanded to a swath of land that is almost the size of Florida. The challenge for managers now is to keep the wolves from attacking the domesticated reindeer that live on the open tundra of northern Scandinavia while allowing at least a few passage from source populations in Finland and Russia to avoid a recurrence of genetic isolation.[5]

Unlike in the wild, the costs of inbreeding can be relatively small in captivity, where the environment is constant and there are some protections from pathogens. There's even a slight chance that systematic inbreeding can purge those rare deleterious alleles that wildlife managers are so afraid of. But it is more likely that these sometimes fatal genotypes will become locked in the population, leaving no chance of escape.[6] The costs of mating with relatives, known as inbreeding depression, rise when a population is stressed—throw in a new disease, raise the temperature a degree or two, or alter the habitat, and reproduction can drop precipitously. It's hard to imagine a more stressful time in recent ecological history—let's say, in the last 25 million years. Habitat fragmentation and rapid climate change will bring out unexpected effects of inbreeding. We will be trucking more and more species around.

I accompanied Darrell Land, a panther biologist with the Florida Fish and Wildlife Conservation Commission, a few years ago on a routine telemetry flight over southwest Florida. Land and his colleagues flew three times a week, 52 weeks a year, to pinpoint the location of about twenty panthers east of Naples. As we waited for a thick fog to lift, I asked Land, who is stout and direct with closely cropped hair, about the restoration project. "In my mind," he said, "it's less about the uniqueness of the subspecies and more about healthy cats." Land acknowledged that this was one of the first attempts to restore an inbred population in the United States, though captive-raised animals were becoming increasingly common.

Peregrines, he said, were successfully reintroduced to the eastern United States starting in 1974. The local birds, known as the "rock peregrine," had already been extirpated. After DDT was banned, a captive stock made up of seven subspecies from four continents was used to re-falcon the Midwest and East. With widely different genetic heritages, the birds have thrived.[7] Some birders protested the artificial introductions and the lack of genetic counseling but, Land remarked, "A peregrine is a peregrine is a peregrine."

Taking off from Naples airport, we left the terra-cotta roofs, turquoise pools, and slate-gray roads behind and flew over a gray-green landscape of swamp forest, pinelands, and hardwood hammocks: core habitat of the Florida panther. O'Brien claimed that it was this land—with its mosquitoes and poisonous snakes—that saved the swamp cat. "A hundred and fifty years ago, the Florida panther had a range that spanned the American Confederacy. But by the time we got to the seventies, everybody thought it was extinct. The only reason it wasn't is that it was hiding in the swamp where nobody went because it was unpleasant."

Aerial surveys have made the search for panthers much easier. The H-shaped antennas strapped to the airplane's struts allowed Land to pinpoint the location of a radio collar across miles of open land. That morning, from the bumpy skies above Fakahatchee Strand and the Florida Panther National Wildlife Refuge and adjacent private ranches, he located 22 panthers. With a tilt of his hand, Land instructed the pilot to circle over each radio signal, and he marked the location on well-worn US Geological Survey topological maps. Once green, the quadrants with favored panther rest areas were now cratered with the black ink of Land's minute scrawl. We circled Panther 83 on the edge of the strand just north of the Tamiami Trail. Panther 59 was resting off the Janes Scenic Trail, east of Picayune Strand. One cat was in a hardwood hammock not far from Wild Cow Island, another within earshot of State Road 832 between Collins Slough and Tom Still Hammock. Almost able to make out the individual palmettos, I hoped to catch a glimpse of tawny coat among the fronds. But by the eighth or ninth panther, I started to dread the 45-degree tip of Land's left hand, the abrupt tilt of the horizon, and the stomach-turning spiral above yet another radio pulse—and there were still many panthers to go before we landed.

Farther north, we circled above forested scraps in the midst of a cattle ranch, the cows and calves lazing in the sun just a hundred yards from a sleeping panther. Ranchland, citrus groves, sugar cane, and truck crops—the tomatoes, cucumbers, red peppers, and other produce that filled salad bowls in winter—did not necessarily exclude big cats.

"There's a magical percentage of native to altered habitat," Land explained. "When it gets too low, cats can't use it, though one female nested in ten acres of cypress, completely surrounded by tomato fields." Frost and disease are even more worrisome. When faced with natural and man-made disasters, farmers can decide to make huge profits by selling their land for golf courses or condominiums.

Our last cat was Panther 99, a young male in distant Corkscrew Swamp. Prior to our flight, Land showed me the skull of one its age, with the broad nose characteristic of the Florida panther and a half-inch puncture just above the eye. The cat bite had been fatal. Most young males would travel far to avoid such encounters, and the Corkscrew ecosystem provided a refuge for adolescent cats. There were no breeding females in the area, so dominant males didn't bother patrolling it. Panther 99 could hunt, unmolested by other panthers, biding his time until he was old enough to strike out for a territory of about 150,000 acres that would have abundant deer and hogs and four or five female cats, the panther equivalent of steak three times a week and a house atop a hill. There wasn't room for many of these territories in south Florida, but even dominant males grow old or die. Panther 99 would be waiting.

The cypress domes gave way to the serpentine twists of new roads. During our flight, we were never more than 50 miles from downtown Naples, yet we circled over 22 panthers. Every biologist I spoke to in Florida turned to a map at some point in the conversation, pointed to the barriers of the panther's range: citrus groves; Big Sugar; the latest development near Fort Myers, Orlando, Tampa, Miami; the hundreds of retirement villages. It was like trying to squeeze a red-cockaded woodpecker into the lots of Boiling Spring Lakes—but a panther couldn't nest in someone's backyard. An increase of 60 cats was critical for the population, but about five hundred people move to the state every day—that's more than twenty people an hour. The state was expected to grow by another three million in the next decade. That's three million more people needing shelter and showers, fresh produce, air-conditioned bedrooms and cars, and three million more desires for McDonald's, Starbucks, Wal-Mart, day golf, night golf, even nature trails—and a whole lot of asphalt to take them to their dreams.

The twenty-year database from these flights directed the acquisition of key areas and the management of public and private lands. During the cool months, a capture team replaces old collars and puts new panthers on the network so that they can be electronically monitored from the air. Roy McBride, a tall and amiable Texan, is responsible for most of these collars. If you want to catch a panther or even a jaguar, McBride is your man. When I visited him on the edge of Big Cypress National Preserve, he met me with a firm handshake and a box of capture records. His six female hounds circled their kennels outside his front door.

When McBride started tracking the Florida panther in the 1970s, he and his hounds often spent weeks looking for a new cat. "Now," he said, "I can't find a place to exercise my dogs without running into one." Although McBride didn't see much behavioral difference in the Floridian offspring of the Texas cougars at first, treed cats were suffering more injuries and escaping more often. What in the end is the difference between a Florida panther and a Texas cougar?

Not much. Both cats are pumas, *Felis concolor,* one of the most widespread carnivore species in the world, found from Tierra del Fuego to the Yukon. Mountain lion, catamount, cougar, painter, panther are just a few of its common names. Although there are a couple of characteristics that have been used to distinguish Florida panthers from other pumas—the Roman nose, white flecks on the neck and shoulders (lately blamed on ticks, not genetics), dorsal cowlick, and crooked tail—before it became isolated in the swamps of south Florida, its range was contiguous with that of the Texas cougar. Historical gene flow, evidence that all North American pumas descended from a common ancestor ten thousand years ago, supported the genetic restoration of the panther. The puma, like the North American cheetah, saber-toothed tiger, and American lion, probably went extinct after Pleistocene glaciers covered the continent, bringing cold temperatures—and humans—over the Beringia Land Bridge. With most of the other large carnivores gone when the ice receded, the puma returned from South America as the dominant feline.[8]

Land gave me directions to Panther 59, one of the first cats located during our flight. On the highway out to Fakahatchee Strand, I counted fourteen wildlife underpasses. The culverts were installed, at a million dollars apiece, to decrease roadkill when Alligator Alley, which sliced through the Everglades, was four-laned in the early 1990s. Though it's rare to hear a biologist praise an interstate, Land was forthright in his support for the changes it brought. Before the underpasses went in, the road was lethal to panthers—collisions with cars were fairly common. Instead of the usual carbon black, one truck's skid mark was brown, Dave Maehr described in his book *The Florida Panther*, "composed of skin and the panther's tawny fur." The cat was "an angry patchwork of exposed muscle, clotted blood, burnt peeled skin, and dirty clumps of hair." And that was a lucky panther, rescued and returned to the wild.[9]

Later, when I explored one of the culverts with refuge biologist Dennis Giardina, we found raccoon, deer, bear, wild turkey, and otter tracks, along with signs of the Florida panther. As we studied the 40-foot thoroughfare, Mercedes, SUVs, Jaguars, pickup trucks, and eighteen-wheelers hurtled overhead at 70 to 80 miles an hour. Despite this success, road mortality remains high. Fishermen sometimes cut holes in the fences to access their favorite spots—holes then used by panthers—and many roads still don't have underpasses. Eleven cats were killed in 2006. The year 2009 was one of the bloodiest on record: Seventeen cats were stuck and killed by vehicles, the last a kitten on New Year's Eve. Three were killed by other panthers, one was shot illegally, and two died from uncertain causes.[10]

To get to Panther 59, I drove the full length of the Janes Memorial Scenic Drive though Fakahatchee. After passing through open prairie, I flushed a dozen wood storks on the edge of a swamp; a mile on, a bull alligator tossed himself into a nearby slough, leaving a younger juvenile and a cottonmouth lazing in the warm white marl. It's a deserted, fascinating road with old logging trams leading off into the wilderness on both sides. The area once held thousands of cypress trees, removed during World War II for PT boats and other military needs. Forty-three species of orchids have been found in the park, including Florida's endemic Fakahatchee ladies' tresses, whose half-inch red flowers were in bloom along the trails.

Panther 59's collar was inactive late that morning, but data from hundreds of telemetry flights showed that a panther's day typically begins around 5:00 p.m., as they hunt for deer and hogs, ready to leap upon a victim's neck as the light begins to soften. I arrived at 4:55 p.m. Number 59 would be just waking up from its nap.

O'Brien's comment on the new vitality of the panther had pleased me when I had heard his talk at the American Museum of Natural History on the genetic rescue. But there the nearest carnivores had been stuffed with straw, their snarls behind plate glass. "The big issue in the West is mountain lions eating people," Dave Maehr had recently told me, "but that's never happened in Florida. In fact, the panther has always been viewed as a sort of demure, laid-back kind of mountain lion." Was that still true now that the cougars had arrived? I urged myself onto the old wreck of a road, longing for that laid-back Southern attitude.

A breeze picked up, and palmetto leaves rustled along the trail. The saw-toothed blades can tear skin, but palmettos provide food and cover for white-tailed deer, black vultures, feral hogs, raccoons, turkeys, and panthers, which use them as retreats to raise their kittens. The royal palms shattered the waning light. With no sign of a panther, I decided to retrace my steps back to the car.

After half a mile or so, I came upon a new impression on the trail: a pawprint. There was a prominent ridge in the moist sand between the pad and the four thumb-size toes. A bobcat, I thought, but when I reached down to measure the track, I felt a tingle on the back of my neck. My field guide identified the three-inch print as that of a panther. It's possible that I missed the track on the way in, but the thought that a large cat had just crossed my path and was now up ahead on the trail gave me a thrill that I have rarely had in my local woods up north. I had better get back to the car, I thought, but not too quickly—not like prey.

Still, the print seemed surprisingly small. I could cover it with my palm.

Early the next week, I asked Land about Panther 59. "He's just one of our cross-eyed, one-testicle, low-sperm-count, inbred males," he told me. Fifty-Nine was pure Florida panther; I could have relaxed—a bit—while moving through his territory.

Back in Naples, I went through his dossier. As it turns out, 59 has two descended testicles and a straight tail. First captured in June 1995 when he was two weeks old, Panther 59 weighed in at a healthy two pounds, five ounces. Two years later, biologist Larry Richardson caught the young male on film: 59 trudged in front of a curtain of saw palmetto, his ears flattened in a wimple of despair, shoulders slumped in defeat. He was probably making a wide detour around an opponent, perhaps an older male in his prime. Claw marks ran from the top of his head down to his eye, and two more scratches circled his ears. Having dropped his game face—the photo was from a camera trap, so no one was around—he looked like a defeated boxer after the dressing-room door closed.

The Texas cougars had just been released. Their whereabouts were followed as closely as an errant spouse's, tracked by a private eye. Here's an excerpt from a 1997 progress report: "On October 13, [Texas cougar 104] was located with male panther #59 northwest of [the Florida Panther National Wildlife Refuge]. On 10 November she was located with TX106 [another Texas transplant], around the Rock Island area. And on 10 and 12 December she was located with male #60 south of the Tractor Camp on the refuge. Nothing is expected to come from the encounters with males #59 and #60 as neither cat is yet sexually mature."

Panther 59 did not keep his head low for long. By the time I went looking for him, he was suspected of having sired at least three litters. He was located that evening, courting a female. Later in the week, the pair crossed the canal from Fakahatchee to Picayune Strand. Although it was drained in the 1960s by the Gulf America Corporation, Picayune is still classic Florida swampland, peddled off in parcels as the Golden Gate Estates, but with no infrastructure—no electricity, sewers, schools, or fire department. In recent years, the state has purchased most of the land, with plans to restore historical water flow and tear up dirt roads, but poaching is still a problem, and the area seems eerily devoid of wildlife. There was tossed wreckage on the old roads, and a few homes off the grid at the edge of the estates. Soon after he arrived at Picayune, 59 moved on to the more productive panther refuge.

I had hoped to join a capture crew during my visit, and spent a few days in Naples waiting for the call. Looking east from a borrowed condo along the beach—between the cumulus clouds and the shadows they cast across concrete roofs, tennis courts, sand traps, shopping malls, and swimming pools—I could make out a thin blue lip of wild land. The panther's presence has helped hold that line.

The call never came. Though I had spent eight months negotiating permission to join the state capture team, in the end, a refuge administrator insisted I would need a federal permit to enter Big Cypress for the last capture of the season. There wasn't enough time to negotiate the byzantine workings of state and federal agencies. If someone had caught me in a camera trap, my face would have shown a look of defeat similar to 59's. So on my last afternoon in southern Florida, I drove myself up to the Caloosahatchee River, the northern border of active panther habitat. In past years, several male panthers had followed a narrow corridor of wax myrtle through ranch land, potato farms, and citrus groves, crossing the Caloosahatchee at the end of an old railway grade.

From Ortona Lock, I followed the waters sloping toward the gulf as I made my way to the right-of-way. I came upon a gate: "Posted Private Property. Hunting, Fishing, Trapping or Trespassing for any purpose is strictly forbidden. Violators Will Be Prosecuted." It was just a few hundred yards to the old railway grade and a short swim across the river. I pondered hopping the gate, then thought better of it.

But young panthers keep going. On one male's three-year journey, he traveled from the panther refuge to the outskirts of Disney World. If he doesn't encounter an available female, a young panther typically makes a huge circle—in this case from the panther refuge to Orlando, Merritt Island, and Vero Beach—part of what Maehr called the "random lifestyles of dispersing males." There might have been enough deer and hogs for him to survive, but after three years of wandering, he never found a female.

Maehr, who died in a plane accident in 2008 while monitoring black bears, was one of the leading contrarians to the government's handling of panthers. He wondered if there weren't a better use for the surplus. "It's a fairly clear prediction that the deer herd should decline dramatically in response to predation, and it should probably drop by an order of magnitude when the next big tropical storm hits and the area floods. It's not a matter of if, it's a matter of when. . . . We should try to take advantage of this high density now." If panthers can be transported from Big Cypress to the Everglades, or from Texas to Florida, why can't a female or two be moved across the Caloosahatchee north to Highlands County?

He had a point. The recovery plan of the federal government called for three self-sustaining panther populations. (The Nature Conservancy's goal, less constrained by political considerations, was for fifteen viable populations.) The administrators were hoping that panthers would cross the river and expand their range, un-aided. Their reluctance to move the cats themselves might have de-scended from a stinging experience in the 1990s, when a few Texas cougars had been released in northern Florida to see how they would fare—but the cougars had been more habituated to humans than the managers had realized. When they started showing up in backyards, the cats had to be rounded up. One of the removed cou-gars ended up in a hunting pen, where animals are fenced in for lazy hunters—a public-relations nightmare for the state wildlife agency, and one from which it still hadn't fully recovered. Maehr thought the managers had better get over it soon: "If the agencies are too spineless to actually move a female or two across the river, then there is some hope that one might do it on its own." But the panther might not be able to wait that long. "A more proactive approach is needed," Maehr warned, "if for no other reason than the landscape is still changing."

* * *

Pick a square kilometer of land anywhere on earth. Odds are good that people live or work on it, and that they'll have easy access to others via road or stream. Except for the north and south poles and a few deserts and forests, the world is covered by road networks and shipping lanes. Chances are high that at nightfall you'll see artificial light emanating from the patch you've chosen. "Nature can be dangerous," Peter Kareiva and colleagues acknowledged in a recent paper in *Science*.[11] So we kill its predators. We suppress wildfires to protect our homes; we build jetties and seawalls to fortify coastlines against the sea. To divert rivers to farms and to keep them from overflowing and flooding homes, we've built so many dams that more than six times as much water is held in storage than runs freely in rivers and streams. People use up almost half of everything that grows on the planet, consuming more than 40 percent of earth's net primary productivity.[12] One study has suggested that less than a fifth of the world's land surface has escaped the direct touch of *Homo sapiens*.[13]

But has it? *Wilderness* has the same Norse roots as "will" or "willful," uncontrollable. *Wild* conveyed the idea of being lost or disordered. One of the earliest uses dates to the eighth century: in Beowulf, *wildēor* refers to the savage beasts of the dismal forests, cliffs, and crags.[14] In the United States, the Wilderness Act of 1964 defined *wilderness* as "an area where the earth and its community of life are untrammeled by man, where man himself is a visitor who does not remain." The act deems at least five thousand acres as the minimum for practical preservation; on protected areas, natural processes continue, but top predators and game are often well managed. All of the whooping cranes of the forty-four-thousand-acre Necedah National Wildlife Refuge were human-raised. Almost all of the panthers on the twenty-six-thousand-acre Florida Panther National Wildlife Refuge have been collared. Many bald eagles have been hacked, or are descendants of hand-reared birds, expanding a population of several hundred in the 1960s to almost ten thousand when the birds were delisted in June 2007.

The paradox: in order to protect our notion of wilderness, we must meddle in it more and more, from the landscape level all the way down to the nucleotides that form each species' DNA. Even genetic rescues aren't new. In 1868 in *The Variation of Animals and Plants under Domestication,* Darwin reflected on an interview with a gamekeeper: "the constant breeding in-and-in is sure to tell to the disadvantage of the whole herd, though it may take a long time to prove it; moreover . . . the introduction of fresh blood has been of the greatest use to deer, both by improving their size and appearance, and particularly by being of service in removing the taint of 'rick-back' . . . to which deer are sometime subject when the blood has not been changed, there can, I think, be no doubt but that a judicious cross with a good stock is of the greatest consequence, and is indeed essential, sooner or later, to the prosperity of every well-ordered park."[15] That "well-ordered park"—about as close as Darwin was likely to get to wilderness in his homeland—is now far grander in scale.

Once every two years, the Fish and Wildlife Service is required to report on its efforts to recover species. In 2006, about one tenth were judged to be improving, a third stable, another third declining, and about a quarter of unknown status. The Endangered Species Act's defenders, who see the glass as half full, note that most listed species whose status is known are either stable or improving.[16] Once protected, animals and plants in the United States are more likely to move from the high-risk endangered status to the lower risk of only threatened. In contrast, 140 mammals, 24 birds, 6 reptiles, and 5 amphibians on the IUCN Red List of Threatened Species deteriorated in status between 2007 and 2008. Only 37 mammals improved during this period, along with 2 birds and 1 amphibian.[17] This pattern has continued for many of the years that the IUCN has been keeping records. One analysis, based on the risks of extinction, found that 262 species would have disappeared in the United States between 1973 and 2003 had the Act not been passed. At the time, 35 listed species had been declared or were presumed extinct. That's 35 too many, but it indicates that the Act might have rescued hundreds of species.[18]

Two thousand seven was a banner year for getting species off the list: the American bald eagle, the Yellowstone grizzly bear, and the Great Lakes gray wolves were delisted, the most in any year since the legislation was passed. (The grizzly bear has since been relisted, and the wolves in the lower 48 have been on and off again over the years.) Brown pelicans, at 620,000 individuals, were delisted in 2009, again thanks to bans on pesticide and hunting.

Success can't be measured by recovery and delisting alone. About 2 percent of the species on the list are presumed extinct. Nine of them have been formally delisted; petitions have been filed to remove several more. For many of these, listing simply came too late. Sampson's pearly mussel hadn't been seen along the Lower Warbash River for 50 years when it was listed as endangered in 1976. After seven years on the list, with range-wide surveys and interviews with clammers throughout its historic range having turned up nothing, it was removed. Two popular fish of the Great Lakes, the blue pike and the longjaw cisco—sold as smoked freshwater herring—had probably been harvested to extinction by the time they joined the Class of '67. They were removed in 1983. The Mariana mallard population may have been beyond rescue when it was listed in 1977. (A hybrid between the mallard and the Australasian black duck, it may have been the youngest species on the list, possibly only ten thousand years old—or not a species at all). Several attempts at captive propagation failed, and the last bird died at Sea World in 1981. It was removed from the list in 2004, along with the tiny Guam broadbill, the last of whose kind was probably eaten by alien brown snakes. The Eskimo curlew and the Caribbean monk seal were both in the Class of '67. The last curlew was shot in 1963, and the Caribbean monk seal hasn't been seen since 1952, when a small colony of seals was recorded on a few coral islands between Jamaica and Nicaragua. The seal was delisted due to extinction in 2008.

Some efforts *were* botched. The dusky seaside sparrow in Florida might have had a chance. The Amistad gambusia existed only in captivity when it was listed; its native spring had been silted over by an impoundment on the Rio Grande. Some in the captive population were eaten by the more common mosquitofish; others mated with them. Seven years after listing, they were gone.

The Act's critics note that more listed species are declining than improving, and only twenty species have recovered and been delisted since the bill was passed in 1973. For those, such as former California congressman Richard Pombo who saw the glass as half empty or even broken, this was proof that the Act needed revision. Few conservationists agreed with his proposed measures to eliminate critical habitat and compensate landowners for thwarted development plans. His legislation passed the House in 2006, but never made it through the Senate.

For many conservationists, there are two major caveats to the Act's success: the budget is too small and the list too short. Funding levels in the early 2000s were about 20 percent of the amount needed to get the job done.[19] Dozens of species have gone extinct while waiting to be listed, with their status sometimes snarled up in delays and legal battles.[20] David Wilcove of Princeton University and Lawrence Master at NatureServe estimated that the actual number of endangered species is at least ten times greater than the current list of 1,900 or so. Most of these organisms haven't been studied well enough, or received enough attention, or had the muscle of legal counsel to get them on the list. At least 42 species have gone extinct while on the waiting list to be listed.[21] Invertebrates, fungi, and marine species remain underrepresented. And, of course, conservation is about more than just avoiding extinction: it aims to maintain healthy ecosystems and keep species common enough to be ecologically relevant.

The central concept of the Act is that we can identify and eliminate threats to these species and remove the species from the list. It is a zero-tolerance law that has prevented the extinction and improved the status of hundreds of species. Managers have the tools—

restrictions on wildlife trade, habitat protection, captive breeding, invasive species management, genetic rescues, and reintroductions to name a few—if not the funding to protect most species. Even for successful recoveries, delisting is a slow process. Some environmental groups and managers have tried to delay the event, fearful of change and loosened restrictions. In 1970, the American and Arctic subspecies of the peregrine were listed as endangered; the bird was extinct east of the Mississippi. Like the bald eagle, it had some help from humans with its comeback: eggs were removed from multiple clutches and the young falcons fed at new hacksites. Peregrines have successfully nested on buildings and bridges from San Diego to Boston, feeding on invasive starlings and pigeons (bless their predatory little hearts) and other urban species. But even after the peregrine was fully recovered and had become a well-adjusted city dweller, it took more than three years to get it off the list.

Species don't get into trouble all of a sudden, Dale Goble, a law professor at the University of Idaho, told me, "Presumably it's going to be a gradual process going [in] the opposite direction. If the human population was reduced to say two hundred individuals, how long would it take the population to reach a thousand, even?" Many listed species are slow breeders. The California condor lays an egg or two every two or three years. If the egg hatches and the chick survives, it doesn't reach sexual maturity for at least a decade. Even when species do start to rebound, Goble continued, "The idea that you could just return the species to some population level and then just walk away isn't going to work for most species." Many are reliant on management actions, such as the control of invasive species or recurring prescribed fires. Goble and colleagues suggest that a new category is needed: conservation-reliant species, part of a continuum between endangered and full recovery. As many as 80 percent of the species on the list may fall into this category.

Goble gave me an example: "The Borax Lake chub is a little bitty fish in a five-acre lake in Eastern Oregon. The entire population of that species is in that lake. It was listed because they were going to do geothermal development in the neighborhood and the person that owned the lake was a rancher who wanted to get water out of the lake to graze cattle. So the land has now been purchased by The Nature Conservancy and the surrounding land is managed by the Bureau of Land Management. It's been withdrawn from geothermal leasing. The area could be sufficiently protected so that it's not clear that they're getting any more protection by being listed." Species like the chub and the Tiburon jewelflower, whose entire range is a third of a square mile on the south-facing slopes of the Tiburon Peninsula in Marin County, California, will never be fully secure.

We have become pastoralists of the wild, arborists of the forests as we've lost species of trees. The transfer of Texas cougars to Florida was controversial, but at least they were pumas, closely related members of the same species. In the race to restore rare plants, molecular biologists have started moving genes between species and even directly from the pathogens themselves into their hosts.

Until the nineteenth century, one of every four trees in the Appalachians was a chestnut. The tree defined the region, turning the mountains into white caps when its creamy-colored catkins bloomed in spring; in fall, they covered the forest floor with nuts as red as fire coals, forming a bed four inches deep. Turkeys, passenger pigeons, elk, and bear thrived on the mast. The nut crop was so thick farmers didn't need pasture land—they notched the ears of their cattle and hogs before turning them loose in the forest to feed. Chestnut-flavored mountain pork hung in smokehouses throughout the region, esteemed by nineteenth-century foodies. After an autumn storm, families raced out to beat the hogs to the crop—an entire winter's worth of food could be gathered in a month. Sacks of chestnuts were hung near the kitchen all winter, to be baked over coals. Some were sold for "shoe money," the earnings spent on children's winter footware.[22]

Commercial timber harvest almost ruined these mountains—at its peak, about four billion board feet were cut from the region, enough to build a plank road 30 feet wide around the equator—but it took a microbe to reduce this mighty tree, the tallest of the eastern hardwoods, to a rare understory shrub. Around 1900, a blight hitchhiked on chestnut trees imported from Japan. Spores were spread by rain-drops and squirrels, the fungus entering the trees through any wounds present, killing the cambium between the bark and heartwood and girdling the tree. The disease was first observed at the Bronx Zoo, where the oldest trees were reduced to stumps. It spread south to Mississippi, north to Maine, throughout the chestnut's range. At least seven native moths have gone extinct since the tree disappeared.[23]

With its tall vase shape, the American elm was the perfect street tree, creating a beautiful dappled canopy in cities and suburbs of the East and Midwest until 1928, when another Asian fungus, this one spread by bark beetles, reached the United States. Dutch elm disease attacked the sapwood of these once-common trees killing more than 70 million. The elm survives in isolated pockets and along the edge of its original range, in Canada and Florida, in Central Park and Tompkins Square. Manhattan's steel and concrete walls formed battlements, protecting the species from attack.

Can genetic engineering restore the chestnut, elm, and other species wiped out by pathogens? It's worked in fruit trees. On Oahu in the 1950s, ring spot virus devastated papaya trees, reaching the Big Island by the 1990s. Broad leaves were reduced to shoestrings, stunting growth. Infected young trees never produced fruit. Genetic researchers injected the gene for a protein in the virus's coat into a line of trees. The new lineage proved resistant to the virus while producing high quality fruit. Transgenic fruit trees, controversial in some circles, are one thing, transferring DNA between species to restore a lost ecosystem quite another. The American Chestnut Foundation has interbred American chestnuts with Japanese and Chinese species, the original (and blight-resistant) sources of the disease. Early attempts failed: the trees looked too Chinese, the leaves too thick and oval, the nuts too large and not sweet enough. The goal is to get a tree that looks like the Appalachian native, but acts as if it has battled the blight for generations.

More controversial is the introduction of bacteria into these trees to help them defend against foreign microbes. *Agrobacterium* can cause tumor-like galls in plants by transferring its own DNA into their roots, but it has also been used to deliver genes into the germ lines of soybeans, sugar beets, and other genetically modified food. Can this pathogen save the chestnut and the elms? An antimicrobial peptide, effective against bacteria, fungi, and viruses, has been inserted into the DNA of the American elm (it's designed to break down in the mammalian digestive system and have little effect on plant growth). Field tests are now comparing these transgenic trees to Liberty elms, which are resistant to Dutch elm disease, and to wild types.[24] The trees with the greatest resistance will win.

To sequester carbon and fight global warming, a genetically modified eucalyptus is now in development—a cold-tolerant "Frankentree," as critics call it. But what about genetically modified species that become naturalized? Should we really release more exotics? The return of native, hybridized chestnuts to the Appalachians could be more palatable. Chestnuts have been irradiated to induce mutations (some of which may fight the blight), and exposed to a strain of the fungus with low virulence to enhance resistance. Efforts to transfer an antifungal gene from wheat into plantlets have also been successful. There's every chance that a strain may emerge that will allow chestnuts to return as a heritage tree and even help reforest mined areas of the Appalachians. They would be a huge carbon sink; they grow faster than oaks, walnuts, and other hardwoods, and they can hold carbon longer than many other fast-growing trees, such as quaking aspen. The goal is to spread the genes to wild trees—or what's left of them anyway. Restoring their ecological role—bringing back Appalachian forests ankle-deep in chestnuts—still seems a long way off.

Evolution soldiers on: In response to fisheries that are managed to take the largest individuals, fish breed at younger ages, spawning fewer, smaller eggs. In dammed watersheds, when fish ladders are too steep for the biggest to climb, trout get smaller. As the planet warms, species migrate earlier or move their ranges toward the poles, if they can. But speciation in the bigger vertebrates is over—our largest protected areas are just too small to foster new species of elephants, rhinoceroses, apes, bears, and cats.[25] So we throw a few new alleles into the genetic monotony of these isolated populations. The species we've chosen to save bear our mark, just as domestic animals do. I think of hand shadows on the wall: An elephant's trunk, made with middle and ring finger. The fingered wings of a bird in flight. The back of a hand arched into a tortoise shell.

What are we conserving when we add new genes? Sure, we are rebalancing an ecosystem, but the Florida cats are now part panther and part *in situ* genetics experiment. (When biologists noted that Texas cougar 101 was more fertile than expected, having raised eight kittens that survived to maturity, they administered a long-term contraceptive to prevent the female from becoming "genetically over represented.") If their numbers don't increase, it's probable that inbreeding will once again take its toll on the panther. More cougars will be needed to restore the genetic health of the population, further eroding its uniqueness. If a panther is a puma is a cougar, then perhaps it doesn't matter very much. Perhaps we should be glad that even a few strands of homegrown Florida DNA from individuals such as Panther 59—battered in youth, victorious in maturity—have persisted. The eastern cougar disappeared from Appalachia and New England more than a century ago. Whatever unique genetic traits it had are gone forever.

10

The days were for working, the nights for driving. I traveled several hours west on 211 from Boiling Spring Lakes in the dark, leaving the coast behind for the Sandhills. In the 1990s, this region had had the second-largest population of red-cockaded woodpeckers in the country, many on private lands. Pine harvest and development were cutting off bird clusters from one another, and the rise of hardwoods was reducing the open longleaf pine habitat that the birds needed to nest and forage. The role of private landowners in these changes was especially worrisome. According to one study, up to thirteen thousand acres of pine forests had been harvested to avoid potential Endangered Species Act regulations. The closer a property was to a cluster, the more likely it would be managed, through shortened timber rotations, to prevent the growth of mature pines—and keep the woodpeckers out.[1]

Michael Bean, then at the Environmental Defense Fund, and several Fish and Wildlife Service biologists thought they knew what the problem was. In 1995, the Supreme Court had upheld the view that "take" included habitat alteration. This was a victory for environmentalists and the Act—the court confirmed that it was illegal to destroy endangered species and their habitat—but it came at a cost. Landowners became fearful; Ralph Costa, who was in charge of woodpecker recovery at the time, was worried that landowners were developing a view that "more is worse." The reward for growing more woodpeckers, or for any endangered species, would be more regulations. Some landowners, eager to avoid future restrictions on their property, were anxious to clear it—the "scorched-earth" policy of several builders.

I sat down for breakfast at the Days Inn in Southern Pines, North Carolina. Three men, dressed in baseball caps and polo shirts, were in the booth in front of me, complaining about the weak coffee and yellowing cream cheese.

"I'll get you some more," the waitress said.

"It's all right," one said, making it clear that it wasn't, and turned back to his mates. " 'If you want to make this a damn bird sanctuary,' I told him, 'you need to buy my land. Then you can have all the birds you want.' "

"They're going to die one way or another."

It became clear that they were talking about a Florida scrub jay, listed as threatened in 1987. They were in town for the golf.

"I asked them, 'Why don't you move them?' They said they can't because there's no more habitat left."

"Why don't you go down there with a couple of friends with a pellet rifle?"

"It's just a little bird."

The waitress came back with some fresh cream cheese.

"That's okay." He waved it away. "I'm done with it. Do you live here?"

"In Southern Pines."

"When did they build all those homes along the course?"

"Oh, they build quite a bit of houses on golf courses around here," the waitress said.

"Why do they build them where I'm golfing?" he asked. "I want to see nature. I want to see it beautiful."

Like it or not, then he might want to see a certain endangered bird. One of the premier golf courses in the country was nearby, a two-thousand-acre resort called the Pinehurst Country Club; it was the site of a novel agreement between private landowners and the federal government, an approach that would transform the Endangered Species Act.

In the early 1990s, Donald Barry, a vice president at the World Wildlife Fund, had penned a warning to a new congressman from the West whose priority had been to gut the Endangered Species Act: the consequences of trying to change the Act in Congress would be the same as "those obtained from waving golf clubs from the top of a barn during a thunder and lightning storm."[2] All attempts to change the Act had been crushed in recent years. But after the 1994 elections, as the spotted owl case was coming before the Supreme Court, things looked very different: Republicans had been swept into power in the House and Senate, and they began an open assault on the Act. After the Supreme Court sided with the Department of the Interior that changing a species' habitat (old-growth forest in the case of the owl) was "take" as defined by the Act, and thus prohibited, private property groups were up in arms. At that time, Barry was deputy assistant secretary at Interior. "We were fighting for our lives," he told me over the phone from his house in Alexandria, Virginia.

In North Carolina, Costa, Bean, and their colleagues wondered if there was a way to provide incentives to landowners in the Sandhills without adding costly prohibitions to the future use of their property. Private land could provide better habitat for endangered species if owners managed it differently. So they came up with a new idea: landowners would sign an agreement with the federal government to abide by the restrictions of the Act, maintaining the current number of red-cockaded woodpeckers on their property. If the owners improved the habitat and grew more woodpeckers, they would only be responsible for the original number of birds on their land. If there were no birds at the time of signing, they wouldn't be hamstrung by regulations down the road. Pinehurst Golf Course signed the first Safe Harbor Agreement in 1995. The birds apparently didn't mind the golf clubs—there are now ten clusters on the property—nor did they interfere with play; Pinehurst hosted the US Open in 1999 and 2005.

"Although the timber industry and the Farm Bureau tried to turn the woodpecker into the Southeast spotted owl, I was committed to not letting that happen," Costa said. The Sandhills region, with high-end horse country, golf courses, a major military base, and tens of thousands of acres in timber rotation, had the red-cockaded's ecology on its side. "You can cut some longleaf pines and benefit woodpeckers. Red-cockadeds actually like chainsaws at some level."

There were skeptics. Conservationists worried that landowners would agree to protect a habitat, only to destroy it later. Landowners worried that new restrictions would be added after they had signed agreements. To counter this fear, Secretary of the Interior Bruce Babbitt devised a new policy: No Surprises. "Most of us thought he had gone off the rails when he first came back and said this is what he wanted to do," Barry told me. Once they signed on, landowners would be assured that the rules of the game wouldn't change. A new discovery or a change in status would not result in escalating regulations.

I remembered shaking my head after Hammond prodded that landowner in Boiling Spring Lakes to get into the program immediately, before she had any birds. "You can't let perfection be the enemy of the good," Barry said. "It is very difficult, if not impossible, to find out what's going on on these private lands, unless you have some reason to believe that something bad is happening. You can see it. You can hear it. But you're like Blanche Dubois in *The Streetcar Named Desire.* You're dependent on the kindness of strangers. . . . You'd never get a chance to see what's really going on unless you had proof positive." As a result, according to Michael Bean, the government and even many environmental groups tended to look the other way.[3]

Safe Harbor and No Surprises provided, in Babbitt's words, "an iron-clad guarantee" to landowners. He also started Candidate Conservation Agreements with Assurances, or CCAAs, which helped property owners protect declining species before they were listed, in exchange for a commitment from the government that no further conservation measures would be imposed if the species was listed later.

Habitat Conservation Plans—modeled on an effort in California to resolve a conflict between developers and the endangered mission blue and callippe silverspot butterflies around San Francisco Bay—gave them flexibility. Developers could plow over some plots with plants that the butterflies depended on for food if other areas were managed to control exotic species and prevent vandalism. Although Congress authorized such permits in 1982, allowing the take of endangered wildlife if it was incidental to a lawful activity and the landowner (or community) entered into a plan to minimize or mitigate the impacts, only fourteen permits were issued in the first decade. Babbitt streamlined the process, and hundreds of plans were issued under his watch, some on areas of more than five hundred thousand acres. Many service employees, like Hammond and Ellis in Boiling Springs Lake, now spend much of their time working on habitat plans.

Essentially prohibitive in its original form, the Endangered Species Act was transformed into a permitting statute. "No" became "if" or "maybe." In *Writing for an Endangered World,* Lawrence Buell, a professor specializing in American culture at Harvard, has pointed out that these new strategies acknowledged the primacy of individual rights over the sacrifices that people were willing to make for nature.[4] Costa told me that many people were happy with the woodpeckers they had, they just didn't want the restrictions that could come with new birds. "Most of them tend to be mom-and-pop woodlot owners, and they only own 70 acres or 100 acres that happens to have a group of woodpeckers on it. Several times, it was the last in the county, just hanging on. It was a big problem for the owners because they typically needed to cut that patch of forest for whatever, medical reasons or sending a kid to school. It was their bank. And when it was time to take your money out of the bank, the guy slinging the paint, marking the trees, sees a hole, and because most of them are honest, he tells the landowner, the landowner calls the Service, and there it starts—the nasty headlines, 'Woodpeckers Devastating Landowners.'" To reverse this trend, programs such as Safe Harbor provide incentives to landowners. The Fish and Wildlife Service gave more than $3.2 million to property owners to help them develop Safe Harbor Agreements for red-cockaded woodpeckers. These changes encouraged landowners to open their properties to federal biologists and manage them differently.

"Babbitt—very consciously, very cynically but, I think, also very brilliantly—said the only way we were going to survive with the Act intact was to divide and conquer people on the Hill who were pushing to gut it," Barry said. Babbitt and his colleagues at Interior approached the timber, homebuilding, and water-development industries individually; they negotiated with large landowners. Soon after the No Surprises policy was announced, the president of Murray Pacific, with more than one hundred thousand acres of old-growth and spotted owl problems, signed on. "All he wanted," said Barry, "was certainty."

Babbitt met with his top-level staff every other week to discuss the Endangered Species Act. "I remember leaning back one moment and had this sort of out-of-body experience," Barry recalled. "As I took myself out of the debate for a moment and looked around the room, I thought, this is pure public policy. That sounds weird, but if you're really into governance, I think people will know what I mean. Nobody was sitting there saying—and I heard a lot of this during the Reagan years and, of course, it's been the hallmark of the Bush administration—'What do the oil companies want?' 'What do these people want?' It was, 'How can we make the Act work?' I found myself thinking that public policy decisions never get any better than this. This is pure heaven."

Within ten years, landowners from North Carolina to Virginia, Florida, and Louisiana had signed up for Safe Harbor Agreements for their red-cockaded woodpeckers. There were 108 agreements in the Sandhills when I visited, protecting more than fifty-one thousand acres and 59 clusters of woodpeckers. It put Costa in a good position. "We still have landowners calling in, wanting to know when they can get their birds," Costa said. "It was one of the hardest things I had to do after Safe Harbor. Some landowner calls up and says, 'I have six hundred acres. Can I have some woodpeckers?'

You have to tell them no." (The recovery plan required new lands to have the space for at least ten groups, or more than 750 acres.) For years, Costa recalled, the worst complaints he got from landowners were, "Come get these damn birds off my property." Now he had to explain why the government wouldn't give them an endangered species.

When I arrived at Julian Johnson's farm, a group of federally listed birds flew over his deck and into his yard. The three landed on a pine just behind his house and nonchalantly started lifting the edges of the bark, eyeing the crevices for ants.

"It used to be that woodpeckers would strike fear in your heart," Johnson said. We watched as the two male red-cockaded woodpeckers foraged on the outer limbs; the female stayed close to the trunk. Thin and clean-shaven, Julian was the fourth-generation of Johnsons in the Sandhills. He wore wireless glasses, a brown fedora, and new Merrell hiking shoes. In the 1990s, Johnson noticed that the woodpeckers were starting to flourish on his land. "We need to do something here," he recalled thinking. "That was still back in the day when it was like, 'You got birds here? Then look out—they're gonna come seize your land.'"

Loblolly and slash pine had been popular crops among landowners in the Sandhills for years: they were fast growing with speedy rotations. Jim Gray, manager of a nearby 3,300-acre farm, who was visiting, said, "Then somebody noticed that pine straw looks pretty good in the yard." It made good mulch: it was lightweight, reduced weeds, and didn't attract termites. By far, the best pine straw was the waxy, bright fourteen-inch needles of the longleaf, which had evolved to protect young trees against fire: the needles burned but protected the stem. Fallen needles brought four to five dollars a bale.

But the oaks had to go, if landowners were to make a profit. "Leafy straw," Gray said, "does not sell as well." It turned out that managing sustainably for pine straw often meant growing woodpeckers.

Johnson offered to show us around. "Some people want the pines strictly for the aesthetics and don't need the income, but I do. My personal favorite is to cut the turkey oaks and black jacks down," he said, "They're usually small. When they start to grow, just spray the sprouts, which saves a lot of chemical compared to spraying everywhere. Of course, your labor cost goes up, but I'd rather spend it on labor than chemical."

We approached a couple of harvesters. The woman had a huge drag rake, three feet wide with six iron tines. The man was shuffling the piles into a baler. Johnson said they could gather about a hundred bales of straw per acre.

He had tried mechanical raking, but the tines pulled out the young seedlings. When properly managed, pine straw is a sustainable crop: it not only protects longleaf pine from being cut down but cypress swamps from dying. A lot of the mulch used in landscaping comes from Louisiana cypress—the shredded bark, not the needles. Once the trees are removed, the swamps become dead zones, vulnerable to hurricanes. By replacing cypress with longleaf, you could protect the bottomlands. Pine straw was looking better and better.

We pitched through the sugar sand in Johnson's SUV, stopping beside a woodpecker hole about fifteen feet above the ground, this one excavated by chainsaw rather than by search and peck in an effort to jumpstart recovery. With resin wells and white paint imitating sap, it looked like the real thing. Woodpeckers had already moved in—the cambium around the nest was bright red from their pecking.

"I was very happy to see the Safe Harbor Agreement come up," Johnson told me as we approached one of his nests. "I think I was one of the first ones to sign up for it. I saw immediately that it was a good way to continue working, have the birds, and not be afraid that we're gonna lose control because of them. It's really a win-win situation, because the birds seem to be doing better and we're doing good with selling pine straw." By making working landscapes available to endangered species, there was a great opportunity to speed up and expand recovery—this was the brilliance of Safe Harbor.

"We've got every age of tree, from grass seedlings all the way to old mature trees growing in the same area," Johnson said. "I like the diversity that this creates. I feel like the wildlife does, too. We have turkey, quail, squirrels, deer. We try to keep everybody in mind with what we're doing."

Bachman's sparrows, redheaded woodpeckers, and brown-headed nuthatches often thrive in areas with red-cockaded woodpeckers, who start cavities that are later used by dozens of species, from bluebirds to rat snakes. There were hoofprints everywhere. Johnson told me that as part of the easement, he allowed the army to do some training on his land, which abuts the southwestern edge of Fort Bragg. Special Forces trained on horseback to prepare for deployment to Afghanistan. They used his hayfields to train parachutists.

"Julian Johnson was our first conservation-easement guy," Ryan Elting of The Nature Conservancy told me. The Conservancy and Fort Bragg had purchased an easement that allowed training and sustainable farming but forbade commercial development. Johnson's property was beautiful, profitable, and patriotic, a thriving forest with both endangered species and Special Forces. He had adapted to the birds and to the new provisions of the Act, and he was what my mother would call a real Southern gentleman. "He's so good," Elting said. "We try not to bother him and bring too many people out here."

As we were getting ready to leave, Johnson told me that he and his wife were getting ready to go to their beach house on the Outer Banks. "That's what the easement paid for," Margaret Johnson said.

Ways we value trees: Once seen as little more than board feet, as goods to be extracted, trees and forests are recognized more and more for their intrinsic values and for the services they provide, the ecological impacts that trickle down to human society at large. The world's ancient forests are a natural way to capture and store carbon, absorbing more than 4.8 billion tons of it each year. Worth about

$200 billion annually, they are bigger and cheaper than industrial solutions to capturing carbon.[5]

Understory vegetation and leaf litter protect the soil from the impact of rain, holding it back against erosion and clearing waterways of sediments. Waterlogged forest soils filter out contaminants, keeping them from reaching groundwater and streams. Trees, especially young trees, are thirsty; they reduce the total amount of water flowing into rivers and streams, the abrupt rises, or flashiness, that can cause flooding. In dry areas, this may not be considered a service, but cutting them down isn't necessarily the answer, either. The loss of tree cover has reduced rainfall in China and may be changing the climate of the Amazon, where three trillion tons of water evaporate from the river each year, later to fall on the continent's breadbaskets in the southeast.[6]

We're not just talking old-growth forests here. A New York City study showed that for every dollar it spends on maintaining and planting street trees, the city earns $5.60 in benefits. They provide an annual benefit of $122 million a year by absorbing CO_2 and providing shade. Wondering if they make a difference in the air you breathe? Trees absorb lung-damaging sulfur dioxide and nitrogen oxides. They absorb ozone and carbon monoxide. An expanse of trees can greatly enhance property values—just look at the price of apartments overlooking Central Park—but even a single tree outside a home can increase its value by about 1 percent. That may not sound like much, but in New York, that's more than $5,000 for a single Norway maple or London plane tree outside the front door.[7]

A landmark study in *Science* revealed the health benefits of simply observing trees out a window. In 1984, Roger Ulrich followed patients recovering from gall bladder surgery in a Pennsylvania hospital. Those in rooms with a view of trees spent fewer days there and used less pain medication. According to the nurses' notes, patients with rooms facing a brick wall were more likely to be upset or crying, requiring the nurses to console them.[8] Those nineteenth-century doctors who sent their consumptive patients to convalesce in the pines may have known what they were doing all along.

As Elting and I drove to Fort Bragg, I didn't know what to expect. How would we get on the base? Would most of the trees be cleared? I had been prepared for days of negotiations to get on the army's property, understanding that, as with the panthers, it might not work out in the end.

We drove right onto the base. There were no checkpoints, no gates. There must have been a sign, but I missed it. Then we headed into one of the most memorable landscapes I would see during my trip through the South. Flowering wild grass was everywhere. Among the thousands of mature longleaf pines, I noticed some of the familiar white bands marking a cavity nest. "This is the most studied population of animals in the world," Elting said. "They know all the parents and offspring for generations."

The older pines grabbed the wind like waves. Long-haired adolescent trees swayed on the sun-dappled hills. The golden wiregrass had a warm, almost inner glow, as if the sun were shining up from the soil below. A beaver pond glistened. Fort Bragg, Elting told me, had more than 1,500 plant species and several rare animals, including northern pine snake, St. Francis' satyr butterfly, Carolina gopher frog, and Sandhills chub.

The US Army's mission is cut and dried: deploy anywhere in the world at any time as directed by the president and Congress. Fight, subdue, or destroy the enemy. There's no talk about woodpeckers or pine snakes. The army hadn't been exactly an enthusiastic party to the Endangered Species Act and woodpecker recovery in the beginning, but it owned lots of longleaf pine. "They wanted a natural area for training," Elting told me, but they didn't want regulations. One training area manager of Forces Command insisted that the woodpeckers "didn't give a rat's ass about training."[9] (The manager admitted this was anecdotal.)

In 1993, the Fish and Wildlife Service, concerned about the base's woodpeckers, had issued a jeopardy opinion on Fort Bragg, which threatened to stop all training. "That would have been a disaster for the Act," said Elting. The Defense Department warned that if endangered species protection threatened military operations, they would seek an exemption. Biologists did not want to argue against the military in the court of public opinion. So a compromise was struck: the army would help buy lands and easements around the fort to supplement the woodpecker population, and training could occur near some of the clusters.

As one environmental consultant told me, "Try to find a wildlife refuge that has a lot of birds. There isn't one. They don't have any money. Military bases have tons of cash." The Department of Defense awarded $7 million to The Nature Conservancy to protect and manage buffer zones around the base. The Conservancy contributed 74 cents to every dollar in the army's conservation program. At least 27 clusters on surrounding properties added to Fort Bragg's recovery numbers.

The North Carolina Sandhills Conservancy Partnership (which included the army, The Nature Conservancy, Fish and Wildlife Service, state and local agencies, real estate agents, and even pine-needle harvesters) managed to achieve recovery for the woodpeckers in 2006, five years earlier than expected. There are now 368 clusters in the fort, up from 238 in 1992. Clusters were succeeding between the firing lines of M-60 machine guns, on tank gunnery ranges, along supply routes, and on private forests like Julian Johnson's.

So how can we move forward with the Act? Most conservationists I spoke to agreed that stronger landowner incentive programs—more funding in the form of tax relief or direct payments to encourage voluntary conservation—were needed, as an estimated 80 percent of threatened and endangered species live on private lands. Incentives would turn owners into partners in conservation rather than its enemies. In 2007, a bill that would have provided more than $400 million in tax credits each year to protect endangered-species habitat or to conduct projects that benefited their recovery was introduced in the Senate. Considered bipartisan and noncontroversial, and passing through committee with no opposition, the proposed legislation appeared to acknowledge the political desire to protect species and the need to balance the costs to all. But it never became law.

Conservation banking, where a landowner can get a premium on land that otherwise has little value for development or agriculture, can actually make the discovery of an endangered species on private property pay. California was the first state to award tradable credits for endangered species—to a six-thousand-acre site that was home to the San Joaquin kit fox, Tipton kangaroo rat, and blunt-nosed leopard lizard. Since then, habitat-protection credits have resulted in valuations of up to $125,000 per conservation-bank acre for rare plant and animal species. (One of the challenges of the credits has been how to quantify them.) When researchers in California contacted such conservation-bank owners, they found that most were in it for the money, and that the majority of the agreements had paid off. As of 2003, the Fish and Wildlife Service had approved 35 conservation banks covering almost forty thousand acres and sheltering 22 protected species.[10]

In Iron County, Ted Toombs of Environmental Defense is working to create credit exchanges for the threatened Utah prairie dog, letting the market set the prices. "We've got prairie dogs here that are living next to people's houses. They're surrounded by buildings, like Wal-Marts on one side and then a housing development on the other. We've got dogs on the golf course. They've got nowhere to move." Negative impacts to these dog towns are offset by funding conservation and management practices on other private lands. "You're purchasing the credits on highly valuable colonies that you know have a long-term chance of persistence," Toombs said, "and you're allowing other colonies that are surrounded by city to be developed on." The builders pay a fee of up to a thousand dollars an acre to mitigate the loss of the threatened rodent. Endangered species are no longer the "handy handle" that can stop a project.

Species banking adds value to lands—sometimes far from urban areas—that harbor endangered species. In an agreement between the Fish and Wildlife Service and International Paper, the company can bank breeding pairs of red-cockaded woodpeckers to offset endangered species requirements in other areas: the company estimated that it could sell these credits for about $100,000 a pair.[11] Areas with high densities of the rare birds have proven more valuable as endangered species habitat than as timberland.

11

The heart of the world lay open, rippling with sunlight. Rick McIntyre and I sat among elk bones, overlooking the confluence of the Lamar River and Soda Butte Creek. It was June, and McIntyre was on a folding chair, long legs outstretched beneath a tripod that held his spotting scope, the valley spread out in front of us like an open textbook. On the right bulged Specimen Ridge, snow snaking down the northern valleys, with white-capped Mount Norris on the left. Big sage covered the hillsides like an old pilled sweater. Down the center, swollen with snowmelt, the Lamar ran steadily, dead trees lay scattered along the bank among live cottonwoods and willows. Along the floodplain, buffalo moved in drifts. The cows were molting, their coats in tatters; on the brightness of the fresh grass, it was the young that stood out, a smoldering ochre. With their snowplow heads and minotaur shoulders, they were improbably beautiful.

Then, down in the sage, something spooked the bison—a black wolf approaching from the east. Two coyotes, lying low, must have caught a whiff as it patrolled the southern bank. When they jumped up and gave chase, the wolf ran to the river, tail tucked deep between its legs.

McIntyre, who has been following the wolves since they first arrived in Yellowstone, told me that this one was Dull Bar, a member of the Druids who has an ash gray bar low on her chest. The pack had formed in 1996 and now included sixteen wolves and six pups. Here was prime real estate, between what's known to local wolf observers as the Picnic Area and the Trash Pullout. The tables and trash were both long gone, but they provided historical landmarks for the · watchers—landmarks that were as bewildering to a newcomer as the classic directions from an old-timer: "Turn left at the old Allen place before it burned down. If you pass where the church used to be, you know you've gone too far." I never sorted it out.

"No one ever expected that the wolves would be so visible," McIntyre told me, his eye to the scope. "They came from an area in Canada where they were trapped and killed. But somehow, on release, they figured out that things were different here."

A female bison stood on the road, looking back at her skittish calf who refused to cross. Traffic started backing up. A few people got out of their cars. Another female approached the asphalt and escorted the calf over the road. For much of the twentieth century, all the bison had had to think about here was the automobile and the odd opportunistic grizzly. The arrival of the Canadian wolves changed everything.

Onto the open plains of the nineteenth century, the once-great hunting grounds of the Sioux, a man led the ox that pulled his cart. From the sagebrush came the startling call of a turkey. The prairie dogs barked at his arrival. A pronghorn kept to the blackened edge of the hill, then disappeared. A hawk covered a hare. He piled the buffalo bones on a cart whose wheels stood as high as his shoulders. By the time he had finished stacking them, a barn owl had taken over the hills. In the dark, the settler made his way to the nearest rail station, where boxcars already overflowing with the moon-colored bones awaited shipment east. Some would go to fertilizer plants; others to the carbon works or sugar refineries. Bones brought about $8 a ton, hoofs and horns up to $15. "Buffalo bones are legal tender in Dodge City," one newspaper quipped.[1] The finest were fashioned into buttons and combs. Hoofs were rendered into glue.

They were the remnants of a passing storm. Several years earlier, hunters and soldiers on the loose after the Civil War had descended upon the plains. The frontier army proved too small and inept to conquer the Plains Indians, but sportsmen and professional hunters learned how to hunt buffalo, the Indians' main prey. Robe traders sent more than a quarter-million hides each year down the Missouri River. It couldn't last. All too soon, the market was reduced to the acres of bones left behind. New settlers living in sod homes and eating boiled weeds, waiting for crops to grow on arid plains prone to thistle-withering droughts, extreme heat, and bitter cold—for many of them, the bones would be the best harvest they ever had.

Bone-picking began when the snow melted in late spring; it closed down with the first snowfall, white lost in white. In the early years, each cart would carry more than half a ton. Great trains pulled long strings of boxcars east. Railway passengers heading west rode through corridors of bone. But load by load, pickers had to travel farther from the stations, farther from the rails, until finally the melting snows revealed no bones—they had all been cleared. It would be the last service the bison would provide.

Once the buffalo were gone, wolf attacks on cattle increased, especially on stray livestock. In retaliation, the Western wolf hunt began, a war fought with guns, leg traps, and poison. Cattlemen baited every carcass they came upon with strychnine sulphate, which killed wolves, coyotes, foxes—any scavenger, really. Left without buffalo, some bone pickers became bounty hunters. It was lucrative for a while: a dollar per coyote, two per wolf in the Montana Territory in the 1880s.[2] The Bureau of Biological Survey, the predecessor of the US Fish and Wildlife Service, got its start in 1905. Its first mission? To eradicate wolves from the western United States. The goal, as one government employee put it, was "to make the finish"—total elimination.[3]

The last two wolves of Yellowstone were killed in 1926—by park rangers, part of whose job description was to kill every wolf, coyote, and cougar they came across. But not everyone agreed with this policy. In 1931, Horace Albright, the Park Service director, declared that national parks should preserve examples of all North American mammals "under natural conditions for the pleasure and education of the visitors and for the purpose of scientific study."

His vision included predators, which had "a real place in nature. . . . [A]ll animal life should be kept inviolate within the parks." Well, perhaps not completely inviolate—predators "found making serious inroads upon herds of game or other mammals needing special protection" would be removed.[4] When wolves returned to the region 60 years later, there would be a similar concern for game animals.

After he left the park service, Albright retreated from his previous stand, concerned that the National Association of Audubon Societies and other organizations were more interested "in saving the predatory species of birds and mammals than giving reasonable consideration to the species that are regarded as very important by the general public."[5] Similar sentiments were voiced across North America. In 1936, the *Calgary Herald* supported efforts to kill mountain lions in Canadian national parks: "There is a pretty general notion that the parks are intended for people. . . . [I]n the capacity of entertainers, the flesh-eaters have certain clearly defined limitations."[6] Shark Week was still several decades away.

As Albright backpedaled in the late 1930s, a park service biologist name Adolph Murie was taking a closer look at coyotes in Yellowstone, then wolves at Mount McKinley. Months of direct observation of these predators showed that they fed largely on the weak and old, often taking animals that were likely to die anyway. His lengthy reports in the Park Service's *Fauna Series* showed that Yellowstone elk populations remained healthy in the presence of coyotes, and Dall sheep persisted in the presence of the McKinley wolves. His endless hours in the field gave him a new take on his study animals: "The strongest impression remaining with me after watching wolves on numerous occasions was their friendliness."[7]

Not long before Murie began enjoying the company of wolves, Aldo Leopold was busy stalking them. A game manager in the Southwest for years, he helped lead the campaign to eradicate predators from Arizona and New Mexico. In 1920, he told the American Game Conference, "It is going to take patience and money to catch the last wolf or lion in New Mexico. . . . No plans for game refuges or regulation of kill will get us anywhere unless these lions are cleared out."[8] The original draft of Leopold's most famous book, then called *Marshland Elegy*, made scant mention of this early career.

After reading it in manuscript, Leopold's former student H. Albert Hochbaum, preparing the illustrations, found the book too filled with regret and self-righteousness. Leopold, highly regarded as a storyteller, had failed to develop enough conflicts. "Because you have added up your sums better than most of us," Hochbaum wrote Leopold, "it is important that you let fall a hint that in the process of reaching the end result of your thinking you have sometimes followed trails like anyone else that led you up the wrong alleys."[9] That hint became one of the centerpieces of the book: a revelation that Leopold claimed occurred with a rifle, recently discharged, in hand. "We reached the old wolf in time to watch a fierce green fire dying in her eyes. I realized then, and have known ever since, that there was something new to me in those eyes—something known only to her and the mountain."[10]

Leopold changed the book's title: *Thinking Like a Mountain* represented the shift in his attention from game animals to all the pieces of the puzzle that make up the ecosystem—though he didn't use that word. He was familiar with the word *ecology*, which had been coined by the nineteenth-century German biologist Ernst Haeckel: the study *(logos)* of household or living relations *(oikos)*. But the term *ecosystem* was still relatively new, first appearing in England in 1935. Leopold's book would eventually have a new illustrator, a larger selection of essays, and yet another title, one that reflected the author's unconditional love for his family's farm—the one we know it by today. A week after he learned that Oxford University Press had accepted his manuscript, he died fighting a grass fire near his shack in Wisconsin. The shack is now a pilgrimage site and study center for conservationists and ecologists around the world.

Around the time that *A Sand County Almanac and Sketches Here and There* was going to press in 1949, Farley Mowat, a platoon commander in World War II and a budding biologist and author, joined the Canadian Wildlife Service. He was sent to Wolf House Bay, northwest of Churchill, Manitoba, to study gray wolves. How much time he spent in the field, and how much of his *Never Cry Wolf: The Amazing True Story of Life among Arctic Wolves* really was true, remains a matter of debate.[11] Like Murie—some say, too much like Murie—Mowat claimed that wolves killed the weak, the wounded, the inferior. He even implied that they survived on small prey such as hares and rodents for long periods of time, an observation later questioned by wolf biologists.[12] McIntyre, a fan of the book, suggests that wolves might snack on rodents like candy bars, but don't survive on them. What isn't debated is the enormous success and influence of *Never Cry Wolf*, which sold more than a million copies. Readers wrote to the Canadian government to protest its wolf eradication program. Disney made the book into a movie. The eastern timber wolf and the red wolf made the Class of '67. The environmental historian Thomas Dunlap noted that, thanks to all this, the wolf, once a pariah, was on its way to becoming a paragon.[13]

The extermination efforts in the West were overwhelmingly successful. Although occasional sightings of wolves had been reported in the northern Rockies, there were no records of reproduction for 50 years. Then, in the 1980s, several wolves crossed the border from Canada. A pack of eight denned in Glacier National Park in April 1986; the first pups were seen in July. Wolves seemed poised to return on their own, until they killed some sheep near the Blackfeet Indian Reservation. "It was a freakin' horror show," Ed Bangs, now the head of wolf recovery at the Fish and Wildlife Service, told me. "The agencies all hated each other. The landowners were so pissed they couldn't see straight." Four wolves were killed illegally, and three brought into captivity. The Service knew it had to do better and appointed a Wolf Management Committee—which recommended reintroducing wolves.

"Congress threw that out on its ear," Bangs said. Even trained biologists were skeptical: the curator of mammals at the Front Royal campus of the National Zoo said, "You hear people talking about putting wolves back in Yellowstone to make it a more natural system. Who are they trying to kid? If wolves ever get reintroduced, it will be in a heavily managed situation. There aren't any gift shops or macadam roads or hotdog stands in a natural system."[14] Bangs was hired to do an Environmental Impact Statement on restoring the animal to the region and began looking in Canada for wolves.

But then the Republican Revolution swept through Congress in 1994. Bangs and his colleagues saw the writing on the wall: "We did the reintroductions kind of in the interim before the new Congress got established. But when those guys got in, some of them went after us, especially in the West, with a vengeance." The chief wolf scientist's position was eliminated; he was shipped to Denver to work on federal aid. Their field biologist was transferred to a fish hatchery. "So we lost three of five people," Bangs said. "And we lost the money."

To tide them over to the next year, Bangs solicited funds from Defenders of Wildlife and other private groups. He was confident that all they needed was a few good wolves. "If I had to," Bangs told me, "I would have fired everybody in the park, including me, to get another batch of wolves in Yellowstone and central Idaho. Because I knew, once you got it started, and it had a good genetic base, the wolves would just take off. It really didn't matter what happened after that."

The wolves brought down from Canada were kept in enclosures in Yellowstone. It was a "soft" release, supplemented with roadkill. At the same time, fifteen wolves were released in Idaho's Bitterroot Range, another twenty a year later. There were no pens for these wolves—once their shipping boxes were opened they were on their own. Both populations have thrived. About 1,700 gray wolves live in the region; and colonizers have spread into Washington, which now has its own wolf conservation plan. More than a hundred sightings occurred in Oregon in 2008. Wolf restoration has been touted as one of the greatest successes of the Endangered Species Act.

"Every wolf does what it wants to do," Rick McIntyre told me as we watched Dull Bar rest on the gravel across the Lamar from the coyotes. She was in no apparent rush to get back to the den. "In most cases, one adult will stay behind to watch the pups, or the pups will be okay by themselves."

McIntyre was out in the park every day before dawn, a latter-day Lorax with a bushy gray mustache and thinning hair the color of fallen leaves. In the summer, that meant rising at 3:40 a.m., watching some CNN, and then driving in from Silver Gate in his taxicab-yellow Xterra SUV, a beacon of wolf activity for anyone in the know. Like the wolves, McIntyre tended toward the crepuscular, staying out only until late morning, when he would announce, "I'm heading in." At his den just outside the park, he would transcribe and enter data. By 7 p.m., he was back in Lamar, sometimes staying well into the dark, long after his spotting scope was of any use.

McIntyre watched a family of tourists get close to one of the wolves. He put on his crossing-guard orange vest. Park managers, no longer mandated to keep predators in check, occasionally have to rope off their celebrity carnivores. On my way into the park from West Yellowstone one morning, I got caught in half an hour of traffic. The cause? Drivers slowing down to watch a bald eagle on the roadside. It wasn't the best place for a nest, but the bird had found an awfully nice snag. Later that afternoon, I got caught in more traffic. After twenty minutes of stop-and-go, I saw the cause: a bull buffalo perfectly happy to use the road to get to a new grazing ground. The two to three thousand bison were once one of the park's primary draws. Fewer visitors thrill to the sight of them these days, but it's the only rubbernecking I can remember ending in a smile. In Yellowstone, there are eagle jams, buffalo jams, hot spring jams, and now wolf jams. The once sleepy Lamar has become an essential stop on tours.

When it was clear that the family wasn't going to interfere with the wolf, McIntyre sat back down at his scope. There were between eleven and thirteen wolf packs in the park, in constant negotiation of their property rights. "It's like the Mafia," McIntyre said. "Wolves are very aggressive in defending their territory. A good percent of adult wolf mortality occurs when packs fight each other over territory. This is the most valuable real estate right here."

McIntyre would know. He started his career as a firefighter, then worked in Denali between 1976 and 1990. When he was there, it was considered one of the best places in the world to see wolves: on a good week, McIntyre might make two or three sightings. He moved on to Glacier National Park before ending up in Yellowstone. As he sits on a crowded ridge, surrounded by newcomers, McIntyre is happy to share his Swarovski spotting scope; each time he made a new sighting, I'd hear him say, "Now, has everyone here seen a wolf?" All were welcome to peer through his lens while the activities of the animal in question were calmly narrated.

"On my first full day back in the park in the spring of 1995, I saw one of the Canadian wolf packs," McIntyre said. For much of the time since then, he has watched them, getting paid for six months, then working for six as a volunteer. As rooted in Lamar as the cottonwoods, he hasn't missed a day since June 11, 2000. When I spoke to him later, I brought up that other stalwart of consecutive attendance. McIntyre said it was his 3,418th consecutive day. "That's about 800 past Cal Ripken. But I'm shooting for ten years. If I include leap years that would be 3,652."

Would he take a day off and leave the park when the day came?

"Why would I go anywhere else?"

Three female elk and their calves caught sight of Dull Bar as she approached the Lamar River. The adults kept their heads high, herding the young through the rushing waters onto a small island.

Black ears and full tail held high, Dull Bar eyed them. It was a couple of hundred yards to the island, plenty of time for the elk to bolt. What to do? She bedded down, white muttonchops almost touching the gravel.

"We saw this one bred by an outside male last year," McIntyre said. "And we later saw that she was pregnant and nursing. Her grandparents on her father's side were among the original wolves brought down from Canada in 1995." A young biologist I had met at the Miner's Saloon in Cooke City told me that McIntyre "eats, drinks, and sleeps wolves." He also thinks wolf, narrating to me what might be going through this two-year-old's mind. "I think it's fair to assume that she's assessing the situation. Seeing what her chances are. If it wasn't for the water, these elk would be charging her. And if there were a few more wolves, she would take advantage of that. One of them would circle back and take the calves."

The coyotes were still on the bank. Dull Bar was restless. One of the most flexible wolves, she was willing to eat ground squirrels should they cross her path; some wolves, McIntyre told me, wouldn't bother with anything so small. She moved behind a small stand of willows. In the distance, a moose browsed among the young trees.

"You talk to old-timers," McIntyre told me, "they said that you'd never see young willows and quaking aspens coming back to Yellowstone." Elk populations in the park had overgrown in the absence of wolves, reaching a peak of about nineteen thousand in the 1990s. They browsed on cottonwoods, on aspens, on willows, reducing the young trees to candlesticks. Streams were exposed, their banks degraded. As the trees disappeared, so did the beavers—they had nothing to eat—and so, too, did their ponds. The elk population now stands at seven thousand. The cause of the decline is still debated. Was it the return of wolves? The hunting of antlerless elk? A long-term drought? Doug Smith, who heads the wolf restoration project, told me, "We were over the carrying capacity of elk. There was really nothing they could do but decline." As Smith pointed out, Yellowstone is predator rich. "We've got wolves, black bears, grizzly bears, coyotes, cougars, and humans." People can't hunt in the park, but seven of the eight elk herds in Yellowstone migrate beyond the park boundaries at some point each year. The age-old question remains: are the wolves increasing mortality or compensating for it by taking individuals that would have died anyway? Observations that elk herds are down where wolves and bears are numerous suggest that these predators are having at least some impact on population size.

No one debates that elk behavior has changed. They were now restless, wary, more elk-like perhaps, with their eyes scanning the valley. Saplings were growing along the riverbanks for the first time in years. The saplings—via the wolves—could help protect Western trout fisheries by cooling the rivers; trout populations had been projected to decline by more than 70 percent in the next 50 years, and early angling closures are a regular occurrence in Yellowstone.[15] Wolves could help mitigate climate change in other ways: as year-round predators, they provide carrion as the winter season declines for scavengers such as bald and golden eagles and grizzly bears.[16] They may even reduce the transmission of chronic wasting disease in their prey.

These are the classic signs of a trophic cascade, when the removal of a species changes the food web and the landscape. Leopold's famous passage predates the term, but beautifully evokes the concept: "I have watched the face of many a new wolfless mountain, and seen the south-facing slopes wrinkle with a maze of new deer trails. I have seen every edible bush and seedling browsed, first to anemic desuetude, and then to death. . . . I now suspect that just as a deer herd lives in mortal fear of its wolves, so does a mountain live in mortal fear of its deer."[17]

There must be a lot of terrified mountains: At this point, only about a fifth of the land on Earth has all of the large mammals it had five centuries ago.[18] Up to 90 percent of the large predatory fish in the ocean have been hunted out.[19] Kelp forests, too, have their mortal fears. After pelt hunters cleared sea otters out of the North Pacific, sea urchin populations boomed. The urchins overgrazed the kelp forest, creating vast barrens.[20] Later studies showed that the loss of otters slowed the growth rate of mussels and barnacles; fish densities declined. Coastal erosion increased, and fisheries and tourism dropped. All for the want of a predator.

Restoration can reverse these cascades. Sea otters have come back along the California coast, promoting kelp growth in areas once overwhelmed by urchins. Wolves are restoring the health of the Greater Yellowstone ecosystem and along the edges of the Great Lakes. Microbial communities change around wolf kills in Michigan, and presumably also around dead elk and bison in Yellowstone. As the microbes and fungi change, so does the nutrient ecology of the forest—plants respond to pulses of nitrogen, creating hotspots of growth.[21]

These changes, although important, are still on a relatively small scale. Wolves in the lower 48 are essentially island dwellers, castaways in a sea of coyotes, who tend to eat smaller prey. Gray wolves are in Yellowstone and the Great Lakes; there are red wolves on the coasts of North Carolina and Florida, though these populations remain small. Bob Wayne at UCLA and colleagues have been looking at canid genetics for more than twenty years. At first, it appeared that the coyotes had swamped the wolves of the Great Lakes area after humans began persecuting them and changing the land. Some argued that this population shouldn't even be listed—they weren't proper wolves. As it turns out, interbreeding has occurred for thousands of year. Coyote DNA—maternally inherited mitochondrial DNA and paternal Y chromosomes—is found throughout the population. Ice Age glaciers may have pushed the wolves south into coyote lands; when the ice retreated, the wolf returned to the Great Lakes a new dog.[22]

Canids are an incredibly fluid bunch. Dull Bar and the other black wolves I was watching at Yellowstone had among their ancestors at least one domestic dog brought to the New World by the first Americans, likely a dog with a black coat, a genetic mutation known as beta defensin, who followed people over the Bering Land Bridge. That coat may be linked to an adaptive advantage in the immune system that helps fight infections; it's been common—especially among forest dwellers—ever since.[23]

One of the park volunteers called McIntyre on the radio: Dull Bar had taken a pronghorn calf early that morning. She wasn't pressed for food. But she hadn't moved for twenty minutes. McIntyre watched as she held close to the river, glancing up at the road between herself and the den. The number of cars along the road was growing as tourists spotted her.

"It's always stressful when wolves are crossing the road," said McIntyre, "but there's a wide range of wariness. It's like the beaks of Darwin's finches" (a large beak is favorable under some conditions, a small one in others), "but here the variation is in wariness. One wolf might be less wary of cars—it can make more trips to the den to feed its pups. Others wait until nightfall. It's better for the pack to have that range. It's true in hunting, too. In most packs, one or two or three adult members in a pack do the hunting. There was probably a similar deal in Ice Age humans. If all everyone did was to go out hunting, then there wouldn't be any division of labor."

Some wolves are born hunters, willing to take on the biggest game. "Like in high school, the most talented athlete wants to play football all the time. But those that are not so good, like me, are happy to play for just five minutes. And I'd probably drop the ball."

McIntyre was an observer by trade and disposition. A flâneur who didn't want to stroll, who wanted to watch not the world going by but the wolf. Spend a couple of days with him, and you could find, as I did, how easy it was to pick up his slow, measured pace and his finish. His jokes were delivered straight-faced, couched in a monotone. But sometimes even he had to move the scene along. Dull Bar approached the edge of the road, eyed the snaking cars, and retreated to the Lamar. From my vantage, I counted 40 cars, most of whom were eager to see a wolf—some, too eager. McIntryre dropped down to the road in his orange vest, tripod on his shoulder, to deal with the traffic. I tried to keep an eye on the elk.

The coyotes were gone. Dull Bar approached the road. Breaking into a run, she crossed and climbed the grassy bank back to the den. When I looked back to the island, I couldn't find the elk calves. I searched the river with my rented scope. Had they been swept away by the river as their mothers patrolled the banks?

The following day, a bison carcass was spotted to the south of Midpoint, a high knoll that provided easy viewing of an open meadow. The Druid pack soon discovered the dead cow and took on what many believe was their true historic role: scavenging. The cow had probably died of natural causes—wolves and bears only rarely take down a full-grown bison; big and aggressive, they guard their young very effectively. "When we see a chase between a grizzly and a buffalo," McIntyre said, "it's usually the buffalo chasing the bear."

The cow's death had orphaned a calf, which was bleating at her side, nudging her to get up. Wolves may ignore adult bison, but a vulnerable calf . . . Dull Bar began an all-out chase. She grabbed it by the neck, while two others began to take it down. The commotion attracted the rest of the bison, who lowered their massive heads at every approach until they succeeded in running off the wolves. Then the cows went to work, licking the bloodied calf until they raised it from the meadow as if from the dead.

As evening gathered, the Druids left the mother's carcass and headed back to the den. A few coyotes moved in. Nighthawks circled overhead. It was too dark to tell for sure, but the calf appeared safely in the herd. In a nearby pond, a boreal chorus frog ticked off time, its call like a fingernail dragged across the teeth of a comb.

Gerlie Weinstein owned the hotel in Cooke City, a few miles east of Yellowstone where I was staying. A friendly innkeeper dressed in purple fleece, she had first come to the park the year after the wolves were introduced, when her son was five. You think, in a park the size of Yellowstone, you're going to see a wolf? Her friends back in New Orleans had been doubtful. "We tried the Lamar Valley. I saw my first wolves at the Footbridge Pullout in 1996. A black and a gray. That was some thrill."

Weinstein was hooked. "I'd come around Mardi Gras and Christmas, because I was teaching. Back then, it was just my son and I, and Rick and the wolf researchers."

She moved up for good a couple of years later and took a chance, purchasing the Alpine Motel. At the time, the Lamar Valley was still pretty remote, much of the tourist trade coming in from West Yellowstone, a more popular entrance to the park. "At first I thought I was the only person in Cooke City that was pro-wolf." But as wolf-watching and snowmobiling took off, the town started to prosper. When the Fish and Wildlife Service first proposed reintroducing wolves to Yellowstone in 1994, it estimated that the carnivores would increase visitor expenditures in the recovery area by about $23 million per year. They were off by more than half: one study estimated that visitors spend over $35 million in the Yellowstone region, just to see and hear wolves.[24] Hundreds of thousands claim to have seen wolves since they were reintroduced. At least one wolf has been observed every day since 2001.

The Fish and Wildlife Service predicted that wolves would take about 19 cattle and 68 sheep each year. In recent years, as wolves have thrived and their numbers increased to more than five times the population originally projected by the Service, depredation has risen: 214 cattle, 721 sheep, 24 dogs, and a couple of llamas and goats were killed in the Northern Rockies in 2009.[25] No one doubts that opportunistic wolves will take livestock, but their impact remains relatively small when compared to disease, or even to coyotes and bears. Local ranchers were compensated about $460,000 for their losses to wolves, a fraction of the millions wolf watchers are pouring into local communities.

These are the trade-offs; wildlife conservation isn't always win-win. Endangered species legislation can bring revenue to service industries like tourism, but that can come at a cost to farming and ranching, a transfer of wealth rather than a loss of it. (Some ranchers have started running wolf tours.) Species protection alone does not appear to limit economic growth: there's no relationship, for example, between the number of listings and any declines in construction jobs or gross state product. In the 1970s and 1980s, Alabama had a booming economy and 70 listed species; Louisiana had only 21 but its economy was ailing.[26]

There are, however, opportunity costs to protection. After the red-legged frog was protected in California, the restriction of pesticides to protect the amphibians could have resulted in about $50 million in crop loss. That 1.6 million acres were declared critical habitat didn't prevent landowners from building on their property, but the costs of delaying construction, as developers await permits, were estimated at between $150 million and $500 million over the next twenty years. Not a huge number when averaged over two decades, but not zero, either.[27]

For the wolves and the ranchers, new methods of predator control may help keep such losses in check. Wolves that attacked livestock were once chased off with rubber bullets; this time around, Bangs and his colleagues insisted that cattle killers got no second chances. They are trapped, hunted—even shot from helicopters. In 2009, 272 wolves were "removed" to control depredation. Livestock losses were originally reimbursed by a compensation fund maintained by the Defenders of Wildlife, which has already paid ranchers more than $1 million; states now cover most of the expenses, including indirect losses. Ranchers claim that cattle raised around wolves, like native elk, have become warier, thinner, fetching lower prices. To help compensate for such changes, Wyoming pays ranchers seven times the replacement value of each calf killed.

Before the wolf was delisted, I stopped at a café in Dubois, a small town on the Wind River southeast of Yellowstone. As it often does in Wyoming, the talk at the nearby table came round to a particular canid. "Wolves are here," said a man in a cowboy hat. (They all had cowboy hats.) "She's got a wolf right here," he said of a neighbor. "She can't go out at night. She has horses and the wolf is prowling the fence line, but she hasn't seen it. The horses are spooked. She's going to have a BYOG party instead of a book group."

"If you don't think you have wolves at your house, think again."

"I love it here," the waitress told me as she delivered my steak. "But it's pretty consistent. I need some inconsistency."

"She called Jackson," said at woman at the table, "and he said she could kill the wolf." At the time, wolves in Dubois, as predators, could be shot on sight, as long as the shooter reported it to a government official within 24 hours.

"If we don't shoot them," the waitress said, "we'll be overrun."

A motherless calf is a short-timer. By six the next morning, the bison herd had moved on. The calf stuck to its dead mother's side, now just a rack of ribs that had been torn open by the Druids, the coyotes, and the ravens. A grizzly worked on what was left of the carcass.

When the Druids arrived, they mobbed the bear, their barks as sharp as hard-driven nails. There were seven pups at the den, and the females were nursing, but still, the chase seemed largely ceremonial. There was hardly enough meat on the bones even for a raven. A new wolf, thick necked and broad bodied, approached from the meadow. The Druids raced him up and down the valley, then left him alone.

When 480, the Druid's 120-pound alpha male, and 302, his older brother, showed up, McIntyre was prepared for the worst. They were the Druid's biggest, baddest males—how would they deal with the newcomer? They barely acknowledged him, tails neutral, almost brushing his side. The new wolf checked out the bison carcass, then bedded down with the pack.

The excitement had attracted most of the wolf-watching regulars. I met a naturalist, a miner, and two former workers from a tire plant. Laurie and Dan Lyman were retired schoolteachers from Southern California. Many of the observers, like the wolves, had gray pelage.

There was a call on the radio. To the west on Jasper Bench, ten wolves from the Slough pack had appeared, making their way through the sage. Someone reported that the wolves had found another carcass just out of sight.

The Yellowstone diehards had their own packlike nature. Everyone was happy to share a scope with a family from Cincinnati. The mom, I noticed, kept to the scope longer than her kids. "I'm just obsessed with this," she said. "When's a good time to see them?"

McIntyre told her that he generally tried to get to Lamar Valley by five.

"I don't do five," her daughter interjected.

When one photographer edged too close to the wolves, tirelessly dogging McIntyre for the best place to perch with his bazooka-size lens, I could see daggers in the eyes of the volunteers. But McIntyre was welcoming. Some individuals, he told me, could move freely about the valley; others were ruthlessly punished, chased off, or worse. He was talking about wolves.

The Druids began a group howl. Then one of the wolves spotted the lone calf, standing vigil not far from its mother's side. The calf ran for a few moments, but without any adults around to protect it, its fate was sealed. There's no easy way to die in the wild. This calf, alone and not yet weaned, would almost certainly have starved; the wolves took it down quickly. 302, who had been the focus of a PBS documentary and had a huge fan base among the observers, worked on the meal for a long time.

With two carcasses to feed on, the Druids could relax. They lazed in the meadow, licking each other, brushing flanks with utmost gentleness. They shared their kill with the newcomer. So why the bad rap? "Until recently, most films of wolves were of captive animals," McIntyre said. "As with people, they get more aggressive when they're incarcerated. When they kill something in Yellowstone, they have no problem letting others eat it."

One morning, I set up my scope on a knoll overlooking the den of the Slough Pack. I had heard that its lone pup was in view, but I was having trouble finding it. One of the early spotters offered to set my sights on the den. It was a long way off, but I thought I could make out a pair of ears poking up from behind a fallen log. We got to talking. Turned out the spotter had recently been working on gut microbes, genetic variation, mountain sheep, and jumping mice. "Jumping mice? Preble's meadow jumping mice?" I asked. Yep.

It was Rob Ramey, blue-eyed, mustachioed bête noir of the conservation world. In 2003, while working at the Denver Museum of Nature and Science, Ramey and his research assistant had compared the tiny, fragile skulls of meadow jumping mice in the museum's collection (gathered from grassy fields along the eastern edge of the Rocky Mountains in Colorado and Wyoming) to related subspecies from Missouri, Nebraska, and New Mexico. The meadow jumping mouse had been fairly common in much of its range, but its dependence on grassy riverine habitat in Wyoming and Colorado—prime commercial, agricultural, and residential real estate—had put it at risk; the Preble's mouse, *Zapus hudsonius preblei,* was listed as threatened in 1998.

The skull measurements matched. Ramey came to the same conclusion on the genetic data: all of the mitochondrial genotypes had been found in related subspecies, and the nuclear data, known as microsatellites, had indicated recent gene flow. Preble's had been interbreeding with its neighbors. Although it had a disjunct range, the rodent appeared to be just another population among millions of meadow jumping mice, a fairly common species complex. Ramey and his colleagues suggested that *preblei* be synonymized with its two closest relatives. The name would be retired.

The listed meadow jumping mouse of Colorado and Wyoming was a strong swimmer with a taste for prime real estate: its riverine habitat was critical to its survival and protected under the law. But Ramey's work suggested that there was no reason that it should stand in the way of developers. The State of Wyoming, which had helped fund the research, challenged the mouse's listing in court even before Ramey's paper had been accepted for publication. Some biologists accused him of being in the pocket of state agencies and real estate developers. One rodent systematist shook his head when I mentioned Ramey. "With enough money, you can always find someone who can get you your answer."

"Subspecies are all over the map in terms of how they are described," Ramey said. (I followed up with him on the phone a few months later.) "They change over time. They're different for different taxonomic groups. It still comes down to, 'A good subspecies is whatever a taxonomist says it is.' The important stuff is the larger issues." He pointed out that more money was spent on the Preble's meadow jumping mouse than on the blue whale and other highly endangered species with longer branches on the evolutionary tree.

"The Zoological Society of London folks get past the problem by saying, 'Here is a priority ranking and we are going to allocate our effort into the top one hundred species.'" The ZSL defines and protects species based on evolutionary distinctness and the level of global endangerment (EDGE). These are the living fossils, often the last remnants of evolutionary radiations. The Yangtze River dolphin, the sole member of the Lipotidae family, tops the mammal list. It hasn't been seen since 1999 and may already be extinct. There are only 2,500 fossas, a lemur-hunter that is the island's largest carnivore, left in Madagascar; none of the island's protected areas is large enough for a sustainable population. Archey's frog of New Zealand, the top amphibian, is so primitive that it doesn't make calls, has no eardrums, and still has tail muscles (but no tail); it has succumbed to chytrid fungus and invasive rats, though a few remain in captivity. The Table Mountain ghost frogs, part of a rare suborder that split off from most modern frogs and toads about 160 million years ago, ranked fifteenth; they are found in a few fast-flowing South African streams, in an area about a tenth the size of Manhattan. When I last looked, the list only included amphibians and mammals; birds were under construction.[28]

Most conservation biologists would agree that it is essential to preserve these unique lineages—"superfreaks," as one journalist called them—as well as those undergoing speciation and adaptation. Shahid Naeem and colleagues have proposed modifying the Endangered Species Act to regulate more than just species numbers: the ecological and evolutionary roles that species play also need protection.[29] One of the big obstacles to incorporating evolution into conservation policy is that the two operate on different timescales. The Endangered Species Act, for example, often judges risk on a scale of ten to twenty generations. Evolutionary change, on the other hand, can proceed at a dinosaur's pace across hundreds and thousands of years or in a single often unpredictable pulse—the arrival of a new species, predation pressure, rising temperatures.

"The Endangered Species Act says 'whatever the cost,'" Ramey noted. The Act elevated populations and subspecies to unwarranted levels of protection. "You could completely drive your nation into economic and social oblivion with that kind of agenda. Without effective prioritization, we're doomed to failure on this venture." I mentioned that scientists like Paul Ehrlich at Stanford argue that the disappearance of a population is often a prelude to species extinction. Ramey laughed: "Ehrlich has been wrong his entire life. Look, he's my great grand-professor. We used to hang out in Costa Rica and drink rum and tonics in the evening and argue. It was huge fun, but if you want to see how wrong Ehrlich was, go read *The Population Bomb*."

Subspecies *are* fluid, perhaps even outdated: morphological variation, geographic distribution, and genetic characters rarely give a clear picture. To many, they are considered units of convenience rather than biology. Some would argue that's the case across the board with traditional taxonomy. A family-level split in one phylum might have occurred at the same time that a genus arose in another. Consider the great apes and fruit flies. The workhorses of the genetics lab, fruit flies date back to the Eocene, more than 33 million years ago; they get a genus designation, *Drosophila*, the second rung on the taxonomic ladder that runs from species all the way up to kingdoms and domains. The great apes, *Pan* and *Gorilla*, evolved in the Pliocene, only 2 to 5 million years ago. Our entire "super family" (slightly higher than the third rung), which includes apes, humans, and gibbons, dates back only about 22 million years. From the point of view of fruit flies, we're all just a bunch of upstart apes: our own genus, *Homo*, has been around for a mere 2.5 million years, and modern humans for a couple of hundred thousand, hardly a blink in one of the eight hundred facets in a *Drosophila*'s compound eye.

Given the constantly shifting, and at times seemingly arbitrary classification system, you could be forgiven if Borges's wonderful "discovery" of the taxonomic division of animals in a lost Chinese encyclopedia came to mind:

(a) those that belong to the emperor; (b) embalmed ones; (c) those that are trained; (d) suckling pigs; (e) mermaids; (f) fabulous ones; (g) stray dogs; (h) those that are included in this classification; (i) those that tremble as if they were mad; (j) innumerable ones; (k) those drawn with a very fine camel's-hair brush; (l) etcetera; (m) those that have just broken the flower vase; (n) those that at a distance resemble flies.[30]

The beauty of this list is how it provides insight—in "one great leap," as Foucault noted—into the limitations, inconsistencies, and occasional biases of classification systems.[31]

How do we address these flaws in the current Linnaean system? It's probably too late to reorder entirely, but we can at least quantify it. John Avise at the University of California, Irvine, has proposed using the fossil record and the molecular clock—the random but relatively steady ticking of mutations in DNA across generations—to add temporal bands to taxonomic ranks.[32] *Drosophila* would be identified with a time clip from the Eocene (33–56 million years ago), chimpanzees and humans with one from the Pliocene (2–5 million years ago). Just as museum plaques help viewers establish the time and place of origin for a work of art, these bands would provide instant and comparable information across taxa.

On the smaller scale, one relevant to management, populations are far more fluid; even the Act's definition of a Distinct Population Segment (DPS)—a term that originated with the Act, not in the scientific literature—has evolved over time. In 1990, a distinct population was defined according to the tenets of evolutionary biology: it should be "reproductively isolated" from other individuals and represent "an important component of the evolutionary legacy of the species." The DPS more or less conformed to the concept of the Evolutionary Significant Unit, first popularized by the zoologist Oliver Ryder: a population that was geographically separated, genetically distinct, and had locally adapted traits.[33] The legal definition has since been modified to reflect political boundaries—a species may be better protected in one country than in another and thus warrant listing where it is most vulnerable; "significance" has been broadened to include unusual ecological settings or potential gaps in the species range. In 2009, when the American dipper was denied protection in South Dakota, federal officials acknowledged that it was "isolated," or distinct, but refused to list it on the grounds that its streamside habitat in the Black Hills was not unique to the region—the bird was still relatively abundant in the Rockies. We have started straying from biological measures here: deciding whether a group is biologically significant or if it represents "an important component" of a species' evolutionary legacy is no mean feat. Some populations belonged to the emperor; others were little more than distant flies.

Most ecologists agree that metapopulations—groups of populations that grow, decline, shoot off new buds, and occasionally blink out—are essential to population persistence. This is the bread and butter of conservation genetics: defining populations and measuring levels of genetic diversity (often a proxy for population health) and migration, or gene flow, between them. Does a species act like one large population constantly swapping genes, or are there several isolated populations, persisting semi-autonomously? At what point is a population distinct?

Biologists at the US Geological Survey responded to Ramey's work with an exhaustive paper that analyzed a bigger sample size, more subspecies of meadow jumping mice, more regions of the genome, and, let's admit it, less restrictive assumptions. They acknowledged that morphological studies failed to identify characters that were diagnostic for the Preble's mouse, but by expanding the number of genes—by increasing the resolution—they found a clear break that supported the distinct character of the Colorado and the Wyoming populations. Instead of a vast homogeneous gene pool, the meadow jumping mouse was justly divided into five subspecies in the southwest portion of the range.[34]

In the end, the Fish and Wildlife Service retained the subspecies designation, but it reassessed the dangers. The Colorado population, which was under greater pressure from development, would be protected, the Wyoming mouse delisted. Colorado's Republican senator, Wayne Allard, heaped scorn on the ruling. "I have a difficult time comprehending how a mouse could nest along Colorado's northern border and wake up one morning listed as an 'endangered species,' cross over into Wyoming to forage for food, and no longer be listed."[35]

Ramey told a reporter that burying the data might have been smarter. "Certainly my life would have been a lot easier if the findings had gone the other way. And I often wish they had. But the truth is they didn't, and I believe that intellectually honesty will achieve more in the long term."[36] Now he has become a lone wolf—his business card was for a consulting company he had started—without the security of tenure in the academic pack.

"By protecting lineages that are actually quite common, even when the original listing was in good faith," he said, "we're spending money in the wrong way. By keeping people away from critical habitats, we could be inhibiting future environmentalists."

The wolf pup made an appearance on the kidney-shaped depression around the den. It rolled on its back, fell off a log. Someone wearing layer upon layer of summer clothes walked up to us. "I snuck out of the hotel room before my husband and kids woke up," she told us. It was the woman from Cincinnati.

Rick McIntyre had almost resented the big, black uncollared male when he first showed up in Soda Butte Valley, at the center of the Druid's territory, on January 17, 2003. At the time, 21 was the alpha male, a strong and stable leader. The new wolf immediately created a sensation among the young females, strolling on the scene and striking a pose, as is the wolf's way. "He was like a biker who breezed into town or the bad boy in high school who all the girls wanted to date," McIntyre said during a late-night drive back from the park. "All the women loved him, even though they knew he was never going to stick around." But he stayed long enough to get a collar and an ID number. "They couldn't get enough of him," McIntyre said. "302 was God's gift to female wolves."

302 had been born into the Leopold pack in 2000, the son of two of the original Canadian wolves released into the park. During my visit, he was eight years old, when most wolves start to show their

age, but he still acted as if he were in his prime. "302 is valuable," McIntyre said, "because he provides an alternative strategy in life. Originally I was on 21's side, but I came around. I'm like 302. I'm not married. I don't have any kids. I'm happy to be a beta male."

"I grew up not far from Harvard," he told me later. "In junior high, we were supposed to write to two colleges to get a catalogue. So I wrote to Harvard and I forget what the other one was. But then I read Thoreau in my junior year of high school and that changed everything. I really had no idea what kind of a major I could go into that would allow me to live a life like Thoreau. The only thing I knew was that at UMass they had a major called forestry; I knew nothing else. I thought, gee, that might be close. . . .

"Things could have easily been different. I could have ended up going to Harvard, and then who knows?" he joked. "Maybe I'd be head of Harvard University Press now. I'd be talking to you about writing the book."

"And you'd be calling me," I said, "asking where it was."

"The real value of 302 is that he showed us how different individual wolves can be, in their personalities, just like people—and, probably, gorillas, chimps, elephants, dolphins, and other intelligent animals. He is a classic trickster, like Jack Nicholson's character in *One Flew Over the Cuckoo's Nest* or Johnny Depp's in *Pirates of the Caribbean*. A rogue, a loner, an agent of chaos who always broke the rules. But he also had the personality type that was very charismatic and very appealing."

And still all wolf. Earlier that week, I had watched as he came upon a pack-mate's regurgitation. He gulped it down and returned to the den.

The battles continued after I left, in the Lamar, in Helena, and in DC. While I was visiting Yellowstone, the Fish and Wildlife Service had been in court—it was always in court—defending its decision to delist the Rocky Mountain wolves in February 2008. Twelve conservation groups sued the federal government, arguing that delisting was premature. The open hunting of wolves, and the genetic isolation of Yellowstone from wolves in Idaho and northern Montana, risked putting the species back in danger of extinction. By the time I returned to Vermont, a judge had overturned the government's removal of the wolves from the list. US District Court Judge Donald Molloy sided with the plaintiffs: "The reduction in the wolf population that will occur as a result of public wolf hunts and [predator] control laws in Idaho, Montana and Wyoming is more than likely to eliminate any chance for genetic exchange to occur." Only wolves caught attacking livestock could be killed. A few weeks later, the delisting of the four thousand wolves of the Great Lakes was also overturned, a judge ruling that recovery should include populations across the historic range of the wolf, not just isolated ones in Michigan, Minnesota, and Wisconsin. The wolves of Idaho and Montana would be delisted again in 2009—and hunting, this time, was opened. Several Yellowstone wolves were shot when they crossed the park boundaries, including 527—a female born into the Druids who had started her own pack and was followed closely by many of the wolf watchers—and her adult daughter. Soon after, Dull Bar moved in and became the alpha female.

The wolf population continued to grow. Montana sold $240,000 worth of wolf-hunting licenses. Ed Bangs told me that, thanks to the hunt, the local view of wolves had "improved dramatically." He was worried that relisting would spark a backlash. "The ESA," Bangs said, "was not designed to manage populations." The wolves were put back on the list by court order in August 2010. Lawsuits continue.

I followed news of the Lamar wolves through Lyman's daily e-mails, sent out to a core group of wolf people. Back at the confluence of the Lamar and Soda Butte Creek, a black yearling wolf had been spotted, eyeing the elk calves still on the island in the stream. What at first had seemed like a good idea—a nursery for the young ones protected by the river—began to show its flaws. The elk cows could go off to graze and return to the island to nurse, keeping a wary eye on the wolves, but the river kept rising as the long winter's snow began to melt: the highest waters anyone had seen in ten years. The elk had been unprepared for the rise: one calf was taken by the Druids, then another a week later; the remaining calf, looking injured, had last been seen heading east with its mother, who had coaxed it across the rushing waters. Whether it survived the Druids, no one knows. While crossing the Lamar, one Druid pup had been swept away from its pack-mates and never seen again.

The new male I had watched approach the bison carcass established his reputation among the wolves and the wolf watchers when he chased two grizzlies off a bison carcass on his own. It seemed that he and 302 would become the bear chasers, helping make carcasses safe for the rest of the pack. But one day he was gone, deciding, perhaps, that his chances of working up to alpha male among the Druids didn't look so good.

When I left the park, it appeared as if 302 would devote his life to romantic adventures. He never challenged his younger brother, who had remained the alpha of the Druids, the one with all the responsibilities. 302 stayed true to his vagabondage, traveling to nearby groups and sniffing out available females. But in the fall of 2008, as he approached the age of nine, he left the Druids with five male yearlings and hooked up with three females from the Agate Creek pack, whose main territory lay to the west of the Druids. In her update, Lyman wondered: "I don't know what 302 thinks he is doing. It must be a late-life crisis." By January, he was sleeker, growing into his new role. He returned to the area of his birth, the Blacktail Plateau. "Of all the male wolves that have ever been born to the Leopold Pack," McIntyre told me, "302 was the least likely to have come back and become the alpha male in his family's old territory. The pups that were born to him this past spring would be the grandchildren of the Canadian wolves that established that territory. He turned out to be a great alpha male and a great father, founding a pack and putting it in the right direction."

By the following October, the Blacktails, as they came to be called, were traveling close to the Quadrant wolves. When McIntyre went down to the area where they had last been seen, he heard a mortality signal coming from 302's radio collar. "That could mean that the wolf has died or sometimes the collar just falls off," he explained. "It will go into the mortality mode after four hours. We were all hoping as we got to the site that it would be a radio collar. It turned out that it was 302. He was dead."

The wolf's body was lying close to the road, but there were no broken bones, so he had probably not been hit by a car. One of the biologists found bite marks around the neck and throat that looked as if they had been done by other wolves, perhaps some members of the Quadrant Pack, who had been seen in the area the day before. 302 looked peaceful. "I'm sure he had been injured many times in his life," McIntyre said. "He probably just thought this was one more time. He would rest and sleep and, when he woke up, feel a bit better." Paw prints around the body indicated that other wolves had come up and sniffed him—probably members of his own pack, as there was no sign of a struggle; the smallest ones were likely from his own pups. Wolf watchers came later to pay their own respects.

302 was nine-and-a-half years old. He was a first-generation wolf, having spent all of his years in the park—where many of the viewers felt that he had changed their lives. His stomach was full of Yellowstone elk meat when he died.

12

I remember exactly where I was on hearing that the Bush administration was going to list the polar bear—I was in my office at the Gund Institute for Ecological Economics working on a paper about whales. It was such a stunning development. The polar bear had been proposed for listing in 2005, but the Fish and Wildlife Service had dragged its feet for almost three years, coming to a decision only under court order. It hadn't been an easy battle: the administration, which had resisted most forms of regulation in general, had in particular fought the listing of species. Though the number of polar bears had increased in recent years even as hunting continued, federal and independent studies were clear: the projected decline in sea ice would threaten its existence.

At the time of the announcement, the CO_2 count in the atmosphere stood at 385 parts per million—before the industrial revolution, it had been below 300 ppm. More than a third of the world's species could disappear as a result of global warming.[1] Would the Endangered Species Act help reduce this threat? Certainly no one who supported the bill in 1973 foresaw that climate change would soon exacerbate, or even outshine the familiar threats of habitat loss, overexploitation, and invasive species.[2] As human populations increased and temperatures rose, would the Act need a complete overhaul to achieve its lofty goals? Secretary of the Interior Dirk Kempthorne began writing rule changes that would prevent the Act from being used against distant power plants. "While the legal standards under the ESA compel me to list the polar bear as threatened," he said, "I want to make clear that this listing will not stop global climate change or prevent any sea ice from melting. Any real solution requires action by all major economies for it to be effective."[3]

The costs of protecting the bear (greenhouse gas emissions aside) would be relatively small. There could be some impact on hunting, and more than 187,000 square miles would be designated critical habitat. (The Department of the Interior estimated that the costs to all businesses, including those drilling for oil and gas, of considering the bear's mostly sea-ice habitat would be a mere $53,900 per year, which would largely be administrative expenses.[4]) The plight of the polar bear would not affect our SUVs, our factories, or our power plants. It would not reverse the steep climb to 385 and beyond. As I had planned my field research for this book, the polar bear had been high on my list. But then I had thought about the irony of flying over thawing permafrost, the real threat to the bears not being a bullet or a leg trap but every mile I spent on a plane. I was no carbon saint: I had flown from Vermont to North Carolina, Tennessee, Florida, Wyoming, and even Africa and China in the past few years. Could such flights be justified? I don't know. Even the Maldives, an island country in the Indian Ocean whose highest peak is just 2.3 meters above sea level, is caught in a dilemma. Carbon emissions are sinking the atoll, but they fuel the economy, the islands being largely dependent on international tourism and fisheries.[5]

Polar bears, like spotted owls, snail darters, and red-cockaded woodpeckers, symbolize their endangered ecosystems. There was always the fear that the bear would bring the Act crashing down around the other listed species, fears that had been expressed for wolves and for owls. The bears could help shift US policy on carbon dioxide or they could reduce the Act to a token gesture, allowing only the meekest of battles to be fought.

According to a blistering review by the inspector general of the Department of the Interior in 1990, 3,600 animals and plants that were known or suspected to be threatened had remained unprotected and largely unexamined. Five hundred and fifty species had been listed since the Act had become law, but that was hardly sufficient. What caused the delay? Listing was salary-intensive—at the time, it cost about $60,000 to prepare a package for a single species. Yet through the years, hours had gone unaccounted for—it was unclear how employees spent their time.[6]

A fatal loophole in the Act's amendments had helped to keep the listing rate dangerously low and recovery plans out of reach for all but the most charismatic and endangered. A species could warrant protection but be precluded because of higher priorities, a deferral that kept a number of them from full protection. At least 34 animals and plants—including the Independence Valley chub of Nevada, 13 Hawaiian plants, the Carolina elktoe mussel, and the Louisiana prairie vole—had gone extinct, lacking protection from the Act.[7] And there was no evidence that the higher-priority cases were benefiting from these deferrals. In short, the Fish and Wildlife Service "had not effectively implemented a domestic endangered species program."

The inspector general recommended the immediate protection of all imperiled species—en masse, using multiple listings to expedite the process. He called for accountability of the expenditures the Service incurred during its status reviews. He estimated recovery costs would be about $4.6 billion over ten years—at a time when the annual budget for endangered species was about $8 million a year. Even with other agencies pitching in, there was a huge gap between what was needed to insure recovery—listing, research, and land acquisition—and the budget that Congress and the agencies allowed. The inspector general's urgent call for the department to survey all candidate species and increase the listing rate by an order of magnitude went largely unheeded.

With biologists all but banished from Washington since the Reagan administration, about the only way to get a species listed was through active citizen enforcement—lawsuits—and after the inspector general's report, there were many of them. More than a third of all species that have received federal protection have litigation to thank—especially three landmark cases in the early 1990s: *Conservation Council for Hawaii v. Lujan* (1990), *California Native Plant Society v. Lujan* (1991), and *Fund for Animals v. Lujan* (1992).[8] The filing of suits didn't slow until 1997, when the Service started using a statutory provision that allowed it to delay the required 90-day findings when they were considered "not practicable." Nothing, it seemed, was practicable, which is why Talbot had expunged the word from the Act years earlier. Between 1997 and 2003, only eleven species were added at the agency's discretion (ten of these under the Clinton administration); 90 percent of recent listings have been the result of lawsuits. The Service contended that such litigation was slowing it down. Even so, in 1998 it petitioned Congress to limit the amount of money available for listing and for critical habitat decisions—a cap that has been renewed ever since.[9]

In March 2008, WildEarth Guardians sought a court order to protect 681 Western species, many of which the nonprofit organization NatureServe considered critically imperiled. Further delays, the group claimed, would violate the Act. Jay Tutchton, their general counsel, had already filed 21 lawsuits against the Bush administration to gain listings and critical habitat—and won every one. The lawsuit on behalf of the Western species dragged on and has yet to be resolved. In 2009, the Obama administration increased the funding for listing to $21 million dollars. (Some put the current cost of a single listing at about $85,000; others suggest it can be as low as $8,000 to $15,000.) The increase was encouraging, but Tutchton said that it would take at least $150 million to work through the backlog, based on the Fish and Wildlife Service's own estimates. Only two new species were listed during the Obama administration's first year.

The administration did endorse a whole-ecosystem approach that encouraged listing multiple species in a single habitat. Earlier concerns that the narrow focus on species might leave endangered habitats behind are being considered. Many biologists had long supported proactive policies to protect conservation areas before species become threatened and endangered. The expansion of incentives such as grants, tax credits, and exemptions from estate taxes could make it easy—and profitable—for landowners to enter into stewardship agreements.[10]

But regulation was still necessary. The courts have been used many times to force the Service to map habitat critical to a species' survival, a requirement of the 1973 law under Section 4. Jim Williams held that critical habitat was one of the standards that couldn't be violated: "If you go back and look at everything that was listed between 1974 and 1985, all the fish and herps had critical habitat. Ken Dodd and I were big believers in critical habitat. Others said, 'Oh, God, I don't want to do critical habitat. It's going to piss people off.' Well, do it anyway. That's your job. Do it."

But don't overdo it. The listing process is supposed to be free of economic considerations, since a species is endangered or it isn't, regardless of business interests, but critical habitat requires balance. Williams told me that at one point, the Service brought in a biologist whose full-time job was to determine critical habitat for species that were grandfathered in. "Unfortunately, he picked some bad examples to start with, one of which was the Sonoran pronghorn antelope." He came up with a proposal of twelve million acres, more than twice the size of Vermont. "It's, like, holy crap! You can't designate twelve million acres of land for critical habitat."

Don Barry told me they needed the designation of critical habitat to get things through. "I felt you needed to have the regulatory gun on your hip, but you wanted to keep it in its holster. The fact that you had it there would get people to comply." So you show up with the gun, but you try not to use it. Barry continued, "I used to joke that a famous conservationist in the twentieth century once said you can get farther with a kind word and a gun than you can with a kind word alone. And I always asked people which conservationist was that? Teddy Roosevelt? John Muir? No, it was Al Capone. That is exactly what the ESA required."

In 1997, the Natural Resources Defense Council sued the Department of the Interior to force it to designate critical habitat for the Cali-

fornia gnatcatcher, which had been listed four years earlier. Only a small percentage of species had critical habitat at the time—the department used a loophole in the law, suspending designation if it wasn't deemed "prudent" or "determinable." Though Congress had intended that such exceptions be rare, they had become a familiar part of the listing process. In the case of the gnatcatcher, the Service argued that publication of habitat maps would lead landowners to clear their property in violation of the law. Yet out of four hundred thousand acres of habitat, only eleven small violations were found, and the US Court of Appeals for the Ninth Circuit rejected the argument. Every critical habitat designation since has been court ordered.[11]

Is habitat designation effective? Species with it do appear to fare better than those without.[12] But part of this may be the benefit of research involved in preparing for litigation. Williams told me that he had always drafted listing packages and critical habitat designations with the courts in mind; it insured that he could defend each decision he made. Others in the Service insist that the great majority of actions that would adversely impact a species' critical habitat would also jeopardize its continued existence, and thus be prohibited anyway. David Wilcove at Princeton has wondered if critical habitat comes at too high a cost: "Private landowners, for whom critical habitat has no bearing, get infuriated when their property is designated critical habitat. So it's a designation that only affects the actions of federal agencies. . . . Much of the protection that critical habitat confers is redundant with other provisions of the Act. So in exchange for angering private landowners and tying the Fish and Wildlife Service in knots—it can't use its limited resources to list more species or engage in recovery actions when it's designating critical habitat—in exchange for that, you get maybe a minor increase in protection. It never struck me as worth the price." Wilcove did acknowledge its usefulness in two contexts: federal management agencies treat critical habitat more respectfully when it comes to the granting of grazing allotments, stocking fish, or other harmful activities, and in court, critical habitat probably weighs on the judge's mind, increasing the likelihood of a favorable opinion.

Lawsuits aren't only about habitat and getting species on the list. In 1999, bald eagle populations had exceeded the recovery goals set by the Fish and Wildlife Service. The Pacific Legal Foundation, a firm based in Sacramento that supports private property rights, sued the Service in 2005 for delaying the delisting process. A district court judge agreed, ordering the agency to finish its proposal. The eagle was declared recovered in 2007.

Early on, Congress and the Fish and Wildlife Service understood that the full impact of the Act could cover thousands of species. Thirty years later the number of deserving animals and plants probably exceeds ten thousand, though the vast majority remains unprotected. In the eyes of conservationists, the federal government has failed to meet the Act's initial goals. Although it stipulates that a species' status should be established "solely on the basis of the best scientific and commercial data available," listing is often a political decision. During his time in office, George W. Bush followed Reagan's lead, placing about 8 domestic species on the endangered list each year. (His father averaged 58 a year, Clinton 63.) A 2009 report from the General Accounting Office confirmed what many officials already knew: the protection of most listed species depended on the knowledge of individual biologists. Since few species had objective databases that monitored cumulative take—the overall impact of federal and private activities— once a biologist overseeing a particular species retired or moved on, institutional knowledge disappeared.[13]

Just as disturbing was the cherry picking of Fish and Wildlife files in the Bush administration. Unsigned notes written by an agency official on May 16, 2005, said that employees could "use info from [internal] files" that refuted petitions "but not anything" that supported them. (Evidence from citizen's petitions was apparently fair game.) The policy was attributed to Douglas Krofta, who headed the Endangered Species Program's listing branch at the time.[14]

The Office of the Inspector General uncovered widespread political meddling in the Endangered Species Program by Deputy Assistant

Secretary Julie MacDonald, a civil engineer with no training in biology. While she had been in charge of listing reviews and critical-habitat designations, the California Farm Bureau, eager to see the threatened delta smelt removed from the list, had submitted a report without references, citations, or scientific research; MacDonald used it to challenge her own biologists' research and peer-reviewed papers, sharing internal documents with the farm bureau's attorney, a vociferous critic of the Act, in an attempt to demonstrate conflict at the Service. (Other documents were forwarded to family, the father of a "friend" she met on an online role-playing game, and accounts ending in chevrontexaco.com.[15]) But none of her efforts could get that rare little fish, clearly endangered by the pumps used to irrigate parts of the Sacramento River Delta, off the list.

In drafting a risk analysis for the greater sage grouse, whose numbers had been declining in the West, she called field staff to bully them into changing their documents. She stated that the cost of critical-habitat designation for the endangered tiger salamander in central California was too high, only to admit later that she had misread the figures on a state website. Despite clear genetic evidence that three populations of the salamander occurred, MacDonald pushed to consolidate them, which would lower their status from endangered to threatened. Her results were challenged and eventually overturned in court. She reduced 296 miles of critical habitat for bull trout on Oregon's Klamath River to just 42. She cut the range of the Southwest willow flycatcher, apparently to keep the bird away from her husband's family ranch. She improperly influenced thirteen of the twenty endangered species decisions made by the time she resigned.[16]

An investigation by the *Washington Post* showed that Vice President Dick Cheney put pressure on Christine Todd Whitman, Bush's first EPA administrator, to ease pollution rules for oil refineries and power plants, an industry that had contributed big money to the 2000 Bush-Cheney campaign. Whitman eventually resigned over the issue; the decision to ease regulations was overturned in court. In a tussle over the release of water in the Klamath River basin, Cheney decided not to approach the Endangered Species Committee—or God Squad—which would surely elicit the wrath of environmentalists and even some moderates. Instead, Cheney and Interior Secretary Gale Norton asked the National Academy of Sciences to review the policies in the region—a risk, as its committees were independent and subject to peer review, but the gamble paid off: the panel expressed uncertainty on the value of higher lake levels for the endangered suckerfish and on the effects of warm-water release on the threatened coho salmon. But after seventy-seven thousand salmon died, followed by a crash of the commercially important chinook, the US Court of Appeals for the Ninth Circuit ruled that the plan violated the ESA.[17]

Bush appointees in the National Oceanic and Atmospheric Administration and National Aeronautics and Space Administration applied similar pressure to their career scientists. When his scientific advisory committee voted 22–2 to tighten Clean Air Bill regulations of particulate matter, Steve Johnson (who took over the EPA after Whitman) had no problem with overriding it. He oversaw the closing of the EPA's library system—which left agency scientists having to beg copies of research papers from colleagues in academic institutions. In an effort to make lasting changes to the Act, Interior Secretary Dirk Kempthorne proposed new regulations to eliminate independent scientific review of federal projects that might affect protected species. Rather than consult with the Fish and Wildlife Service or NOAA, federal agencies such as the Defense Department, Forest Service, and Department of Agriculture would make their own determinations as to whether their actions posed a risk to a listed species or critical habitat.

But they didn't have a great track record in this regard, having objected to restrictions for everything from red-cockaded woodpeckers to spotted owls to invertebrates dependent on the Edwards Aquifer in Texas. The Fish and Wildlife Service would be left with little, if any, say over the flow of water, a military maneuver, or the extension of an interstate highway. It would be prohibited from considering greenhouse gases and their impact on endangered species in reviewing new projects such as mines or oil fields. The impact of global warming on polar bears and corals, for example, would not be allowed to affect decisions on a particular action. These rules were finalized six weeks before Obama took office.[18]

Left with nowhere else to turn, the environmental community responded to many of these changes with litigation, setting the Bush administration on a collision course with the judicial branch. After the agency delisted the entire bald eagle species in 2007, a district judge in Phoenix ordered the Department of the Interior to redesignate the ones in the Sonoran desert of Arizona as threatened; the department had ignored, then buried, the findings by a scientific panel that the Arizona population was too small to be viable.[19] In the case of the Klamath River's salmon, it was the fishermen who sued—a brilliant move, as that constituency isn't considered at odds, at least not usually, with farming and ranching.

The Obama administration reinstated the independent scientific reviews of new federal projects but opted to let the greenhouse-gas policy stand in regard to the polar bear, arguing—as the previous administration had—that the Endangered Species Act wasn't the right law to use against climate change. Still, the threats were clear: the low-lying northwestern Hawaiian Islands could lose 75 percent of their land to rising seas; Hawaiian monk seals, green sea turtles, the Laysan finch, and other protected species that nest, nurse, and feed there could disappear.[20] Changing rain patterns in the southern United States and reduced snowmelt in the West will only heighten the conflicts between water use and the protection of fish, mussels, and other riverine species. So climate disruption rides alongside E. O. Wilson's four mindless horsemen of the environmental apocalypse: overexploitation, habitat destruction, invasive species, and the spread of disease.[21] Or global warming overtakes them in the coming decades when, if he is right, the extinction rate increases by an order of magnitude.

In 1989, as the environmental movement floundered, the Center for Biological Diversity was founded in Tucson by a bunch of Earth First!ers. The CBD's earliest victory was a lawsuit against the Fish and Wildlife Service to reintroduce an experimental population of the Mexican wolf to New Mexico and Arizona. It became one of the leaders in litigation, going on to file injunctions against pesticides and pipelines that would put species at risk and to petition for listing hundreds of species, including the Kittlitz's murrelet (2001), the staghorn and elkhorn corals of the Caribbean (2004), the polar bear (2005), and twelve of the world's penguin species (2006). In 2007, it petitioned to add the ribbon seal and the American pika—or boulder bunny—that lives in the mountains of the West and can die from overheating. In 2008, the Pacific walrus was added to the growing list of species at risk from climate change that the center proposed for protection under the Endangered Species Act.

The CBD, which has won more than 90 percent of its lawsuits on behalf of endangered species and critical habitat; Environmental Defense; Natural Resources Defense Council; and WildEarth Guardians have argued for hundreds of plants, invertebrates, and other poorly studied organisms, and led many of the legal battles over high-profile species. The role of wildlife groups in protecting endangered species cannot be overstated—and is not restricted to legal action: my trip through the South was signposted by The Nature Conservancy, whose land purchases take into account endangered species and their habitats. Its nonconfrontational oakleaf showed up everywhere from coastal North Carolina to Arkansas. There are thousands of other groups, with causes as broad as Wilderness (Society, 1935) and Wildlife (Defenders of, 1947) and as narrow as a single taxon: Society of Tympanuchus Cupido Pinnatus, which helped save the Wisconsin prairie chicken from extirpation (1961); the North American Bluebird Association (1978); The Desert Tortoise Preservation Committee (1974); and Polar Bear International (1989). Perhaps the oldest, the Sierra Club, founded by Muir in 1892, proved slow growing at first—though there were only 663 members after the first ten years, it now boasts more than a million.

Can the plight of a single white bear change the trajectory of American consumption? It's unlikely that the listing will have the impact of early endangered-species legislation. The more we understand what climate change can do to us and to other species, the more worrying it becomes. It will not be gradual. It could be abrupt and violent. As John Waugh, then at the IUCN, told me, "I'm not sure how much the ESA listing can help at this late date. I spent several years fighting forest fires out West; when a fire crosses a certain threshold, there's literally nothing you can do but step back. For all your efforts, you might as well just throw rocks at it. I fear that in the case of the polar bear and the Arctic, we may be looking at similar thresholds." And then came his rejoinder: "We still have to try."

The polar bear may be the mascot for the perils of global warming, but almost all endangered species stand to be further imperiled by climate shifts. Common species may be threatened by the creation of novel climates and new "non-analog" ecosystems (ones with no prototype); existing ecosystems could disappear from almost half the world.[22] Yellowstone populations of the polar bear's closest relative, the grizzly, could see sharp declines in an important food source, the white-bark pine, as lethal outbreaks of mountain pine beetles, blister rust, and the frequency of forest fires increase. In 2008, the Natural Resources Defense Council petitioned to list the tree, a keystone species that regulates runoff and erosion.

You get the sinking feeling that someday almost every organism will be on the list. In a controversial paper, Chris Thomas at the University of Leeds estimated that between 15 and 37 percent of species will be "committed to extinction" by 2050 because of climate change.[23] Cagan Sekercioglu and colleagues at Stanford and Duke modeled the effects of elevational limits on land birds; their best guess was that, as a result of surface warming and habitat loss, 400 to 550 land birds will go extinct in the next century. Species that live on mountains and in tropical forests may be hardest hit. An animal or plant that ranges across a thousand vertical meters of a mountain range will be forced to move up the slope. As temperatures rise, and the species moves uphill to avoid overheating or pathogens, it will be marooned on the peaks, its habitat reduced to a few cool "sky islands." Different species travel at different rates: birds and butterflies may move ahead of plants and trees, tearing ecological communities apart. No matter. Once a species retreats to the top of a mountain with nowhere left to climb, there's only one step left: extinction. Here in the Western Hemisphere, where two-fifths of the world's birds reside, one degree of warming by 2100 wouldn't be so bad, killing off only about one in every hundred species. From Death Valley in California, 86 meters below sea level, to Mount Acongagua, rising 6,959 meters in the Chilean Andes, a worst-case scenario of an increase of 6.4°C would remove 30 percent of all birds—not just the ones we're used to thinking of as rare.[24] For managers of endangered species, such changes may require that they focus on the highest edges of the range (in altitude and latitude) as species attempt to shift their ranges to cooler climates.

The crop-destroying, lake-emptying drought in the Southeast in 2007 was probably just the first of many prolonged dry seasons, one of the consequences of global warming in the United States. The number of wildfires is likely to rise, snowpack in the Rockies shrink, reservoirs in the Southwest be left with little reserve. Warmer tropical seas mean mass extinctions—that's what researchers have found in the fossil record. Many of them fear that such an event may occur over decades rather than millions of years, unless greenhouse-gas emissions are curbed.

Under this enormous threat, there is an opportunity for conservationists. Ecological economics has made it clear that we can't grow beyond the carrying capacity of Earth without dire consequences. When the governor of Georgia was faced with a drought, he prayed on the steps of the capitol building in Atlanta. The best response may be to do everything we can for the polar bear, the freshwater mussel, the endangered scrub mint, and the whale. To the many ethical reasons to protect all the species on Earth, climate change has added a practical one: protecting biodiversity will store and absorb CO_2. Canada's boreal woods absorb almost a quarter of all of the carbon locked away in terrestrial systems each year. Provincial governments and aboriginal leaders have obligingly set aside vast tracks of trees, wetlands, peat, and tundra, protecting more than 250 million acres from logging, mining, or drilling.

When the Coalition for Rainforest Nations put forth a plan in 2005 that would allow wealthy nations to mitigate climate change by reducing emissions from deforestation and habitat degradation (REDD), there seemed a chance that carbon offsets would be key to saving biodiversity hotspots. If such programs work, they could change the calculations of conservation in tropical countries, supporting biodiversity, climate stability, and sustainable development while alleviating poverty. Among the least controversial documents presented during the failed talks in Copenhagen in 2009 was a new and improved version of the forest plan, REDD+. The agreement encourages tropical nations to protect their forests and the climate while enhancing carbon stocks on working timberlands. We can restore forests, prairies, and wetlands—and pay locals to steward their environment. By restoring and conserving natural infrastructure—by practicing what Michael Rosenzweig calls reconciliation ecology, in which people and species share the working landscape—we can create jobs and provide ecosystem services to the most vulnerable populations, dependent on the forests, oceans, and their own crops for food.[25] And perhaps we can avoid some of the billions of dollars in damages predicted to occur under business as usual.

13

I stood on the bow of the *Nereid,* a 27-foot research vessel. With no whales in sight on the choppy sea, my mind would wander to what the Gulf of Maine would have been like five hundred years ago, before commercial whaling began. Hundreds of right whales were probably feeding on minute copepods, leaving their bushy V-shaped blows at the surface. There would have been finbacks, humpbacks, minkes—and maybe, just maybe, an occasional gray. The disappearance of the gray whale from the Atlantic remains a mystery. Was it hunted to extinction? Had it already disappeared before humans took to the sea with lances and harpoons?

There was a slick on the chop, and then the enormous head of the first right whale broke through. Right whales are incredibly buoyant; that they floated after death made them more attractive to whalers—made them the "right" whale to kill and now among the most endangered. Some rose with their rostrums covered in mud from a deep foraging dive in search of large patches of zooplankton. Before they fluked, a few mud-brown logs were released at the surface: whale turds. They floated out of view.

Several months later, in a mammalogy class at the University of Florida, John Eisenberg was lecturing about sloths. About once a week they descend to the forest floor to defecate, he told us. That's a big deal for these folivores, an hour-long journey from the rich, safe canopy down slowly—painfully slowly from a human's point of view—to the forest floor. On the ground, sloths without tails leave their feces on top of the leaf litter; those with tails punch a hole in the litter to deposit theirs. Beyond the usual blood-sucking flies, ticks, and lice, sloths host a large community of arthropods, many of them sloth-specific. Close to a thousand scarab beetles have been found in the fur of a single sloth, clustered around the elbows and knees. Adult moths spend most of their lives in sloth fur, hiding from bird predators and surviving on sloughed skin and the algae that grows on the hair. Many of these also spend part of their lives in the dung at the base of trees. Beetles and moths deposit their eggs in the dung; larvae feed exclusively on the feces.[1]

The motivation behind the vertical movement in sloths remains a mystery. Were they marking their territory? Unlikely, as they're arboreal, and an errant sloth wasn't likely to come across the marking. Or were they fertilizing the base of the trees they depended on?

Eisenberg mentioned that studies of salmon and bears show that grizzlies play a role in dispersing marine nutrients into the forests surrounding salmon streams. When the fish return to their natal streams to spawn, most die, releasing nitrogen into the waterways and thus to riverine plants and trees. Bears prey on them and then spread the nutrients even farther when they defecate and pee. About a sixth of all the nitrogen found in spruce trees surrounding salmon streams comes from the sea; bears release the great majority of it.[2]

Later, at a biker bar off campus, where my advisor Brian Bowen held informal lab meetings late into the night, an idea floated up through the beery haze: if sloths move down through the canopy to defecate, what about those whales diving for energy-rich crustaceans, then rising from the depths to breathe, and poop?

The classic story in the ocean is one of sinking. In many areas of the Gulf of Maine, nitrogen levels at the surface are so low in the summer that they approach zero, limiting the growth of phytoplankton. Copepods—the right whales' primary prey—and other zooplankton often feed on phytoplankton along the surface at night, then migrate down the water column to escape predators by day. When they go deep, the ammonia they excrete takes nitrogen away from the surface. Their fecal pellets sink. Their own deaths take nitrogen, phosphorous, carbon, and iron away from the surface layer, reducing primary productivity. As it is too dark at the bottom for phytoplankton to grow, the nutrients are considered lost. This pattern is known as the *biological pump*, as if all living beings contribute only to a downward flow.

But watch a whale long enough, and you'll see a different pattern. Many whales feed at depth and poop at the surface. (In case you were wondering, right whales produce brown or red logs, which float at the surface before breaking up. Humpbacks and many other fish-eating whales tend to release broad plumes.) Their upward movement is obligatory—they have to come to the surface to breathe. By releasing nutrients there, they could be creating a *whale pump*. But did they transfer enough nitrogen to make a difference? I worked with Jim McCarthy, a biological oceanographer at Harvard, to build the model. Our work showed that whales, along with other air-breathing vertebrates such as seals and seabirds, transfer thousands of tons of nitrogen to the surface in areas where they feed: they are, in a sense, fertilizing their own garden, bringing more nitrogen into those areas than all rivers in the region combined do.

A few researchers welcomed the idea. Marine mammalogist Sam Rigdway and a colleague had written in the 1980s that cetaceans could lift nutrients from deep waters, in a process that resembled oceanographic upwelling.[3] He told me that when he had watched dolphins from an underwater acrylic chamber in the Pacific their feces came out and disappeared "in a cloud within a very few meters and very few seconds." The nutrients appeared to be released immediately, close to the surface.

Others resisted the concept, suggesting that large-bodied and relatively rare animals couldn't have much impact on ocean productivity. And there were devastating implications: as marine mammals have recovered from being overhunted, some countries have insisted that whales and other predators should be culled to reduce competition with human fisheries. This position is championed by the Japanese government, in part to justify its "scientific" whaling program and resume commercial whaling: If whales eat "our" fish, the thought goes, then killing them is an efficient way to protect fisheries and harvest some high-priced *kujira,* or whale meat, in the process. Several recent studies have shown that marine mammals have a negligible effect on fisheries.[4] And the whale pump hypothesis suggested that cetaceans actually increased productivity in areas where they feed. The relationship between whales and their prey was far more complicated than whalers would have you believe.

One of our reviewers had had a good point: there were feeding aggregations of whales not far from the lab at Harvard where we had done some analyses—why hadn't we gone out there and tested our hypothesis? I emailed Mason Weinrich, who has studied humpbacks off the coast of Massachusetts for years, asking if he had any humpback poop available. Within a few minutes, he replied: "I have several samples sitting on my desk, actually—preserved in alcohol—and we carry a 'pooper scooper' net with us wherever we go." When did I need them?

Great. But the trouble with analyzing ammonium is that you really need fresh feces. How quickly does the dung break down? How long does it stay at the surface? Do phytoplankton use it? I'd have to go to sea to find out. After a few months of discussions, Dave Wiley, a whale biologist on Stellwagen Bank off the coast of Massachusetts, offered me a berth on a 187-foot research vessel. As principle investigator on a project to learn everything about humpback whales— Where do they feed? What do they eat? How much time do they spend in busy, risky shipping lanes?—he thought our nutrient work, while admittedly quirky, would complement his own research. One of the great joys of science has to be turning a thought that surfaced one night over a few beers into a full-blown field project.

In July, I boarded the *Nancy Foster* on its first leg, up from Woods Hole to Stellwagen Bank, an underwater plateau north of Cape Cod, where every summer several hundred humpbacks come to feed. Also aboard the ship were a group of scientists, 2 whale observers, and a crew of 22.

Most of the biologists were there to deploy and then analyze data from a couple of DTags, digital acoustic recorders that would be attached to the backs of whales. We launched the small inflatable boats into what Jeremy Winn, boat driver and keeper of the tags, described as "whale soup"—at least a dozen endangered humpbacks within a few hundred yards, their blows surrounding us like smoke signals. A few appeared to be sleeping, their dorsal fins just above the surface.

Winn, Cara Pekarcik, Becky Woodward, Mason Weinrich, and I set off aboard the *Baleana,* a white inflatable boat, for a group of three whales. Pekarcik soon identified them as Fern with a calf and a female associate, Glo-stick. We watched a female lift her chin, exposing the armored, mace-like bumps on her bonnet, then her dorsal fin and the bony ridge of the vertebrae. Finally, her tail splashed the surface. The big white wings of her flippers spread out beyond the pontoons of the boat. The sea lay calm, dappled with ripples. Cape Cod was a comb-over in the distance.

On the smaller *Wavelet* was Ari Friedlaender of the Duke University Marine Lab, long hair tied back in a ponytail, the number of his latest tag in black Sharpie on his inner forearm: 211-4hr. 149.0334. Of all the ways to study whales, setting a tag probably comes closest to the skills once required of hand-whalers. You had to know when the whale would surface, how it moved on the water, and how to hit it just right before it fluked. So a good tagger was in high demand, just as an experienced boatsteerer or harpooner was in the nineteenth century. Friedlaender had traveled from Stellwagen to Norway (killer whales) to Hawaii (beaked and pilot whales), to more humpbacks on their breeding ground off Bermuda, to the Antarctic, where the whales were so fat and lazy that ice crusted over their backs during the day.

Woodward set up the tag pole, a 30-foot carbon rod balanced on a cantilever. An engineer by training, she did the radio check, holding the tag like a very expensive bath toy. Four suction cups made of medical-grade silica—the first cups used had been designed for roof racks on cars and had proved a little stiff—would be the only parts that would touch the whale. The tag encased an electronic block the shape of a cell phone. Woodward placed it on a small hand-held robot at the end of the pole, which bounced like an enormous fishing rod even in the calm. To release the tag, the timing had to be impeccable, the strike accurate—high on the back so that we could follow it with a radio transmitter.

Woodward stood on a small deck, Winn at the wheel, our modern-day boatsteerer in a faded green New Bedford Whaling Museum cap. Weinrich was at the stern, tall with a salt-and-pepper beard. He watched the whales, following their trajectory even when they were below the surface, out of view. "You've got Glo-stick right in front of your bow," he told Winn. "You've got a fluke print. The calf is a little bit to the left of the bow."

Winn eased in. Woodward flipped the pole so the suction cups were facing down toward Glo-stick.

"That is our whale," Weinrich said. Though humpbacks can be distinguished by their fluke patterns, which range from jet black to pearly white with endless Rorschach variations in between, when they don't show their tails, they can be difficult to identify. This was where Weinrich and Pekarcik came in. They seemed to know everyone at a glance, recognizing them by flukes, dorsal fins, or by the way they moved at the surface. One whale mooed. "That must be Alpha," Pekarcik said without looking up from her notes. "There's only one whale that breathes like that." It was as if Weinrich and Pekarcik were hosting their own humpback party. (I had had the same feeling, watching Yellowstone wolves with McIntyre.)

When Glo-stick resurfaced, Winn and Woodward—who are married—had a second, maybe two, to attach and release the tag to the whale's dark skin. It was a near-perfect deployment, a few feet back from the blowholes, just slightly to the left of the vertebrae.

As the whale descended, it was as if we were holding on to it, the hand-size tag measuring pitch, roll, heading, and depth. The ocean surface dulled into hammered steel as the clouds moved in. We watched as Fern, a protective humpback mom, sank and resurfaced, maneuvering to stay between Glo-stick and her own calf. The tag, designed by Mark Johnson and Peter Tyack at Woods Hole Oceanographic Institute, held two hydrophones and a set of orientation sensors. Set to burn out at sundown in ten hours, it collected a dense data set of the whale's every move and sound, every second of it, for the rest of the day.

On the horizon, a container ship made its way into Boston Harbor. For years, the shipping channel had gone through the productive feeding grounds of humpbacks and right whales, putting dozens of them at risk of being run over. For years, Wiley and his colleagues had collected data, showing the shipping patterns and where and when the whales fed. The ships passed right through some of the densest feeding areas. Rather than bringing his work to a government agency, Wiley took his charts and graphs directly to the shipping companies. "We showed them that by moving the channel slightly to the north," Wiley said, "we could avoid potential collisions." It was a straightforward argument—it wouldn't cost all that much in fuel or time. With the shippers on board, it was easy to persuade the Coast Guard and other federal agencies to support the idea; the International Maritime Organization confirmed the move, and the lane was shifted in 2007. Whale collisions have decreased.

Back on the *Baleana*, we were in the midst of feeding whales. "Don't go into the bubbles," Pekarcik warned Winn as an inverted rotunda of bubble rushed to the surface, "or we won't be happy. Neither will it. Shit! Pull back."

An enormous mouth, dark and slick, erupted. A theater curtain of white rorqual eclipsed the horizon, the pleats open wide, exposing pink ventral grooves. We were so close, I could make out each of the individual barnacles that lined the edge of the whale's mouth. Patches of dark pink flashed through gray baleen. As the whale closed its mouth, several sand lance—pencil thin with long bottom jaws— desperately lunged from the slab of water spilling over the lips. The mouth was so cavernous—a humpback can hold more than 2,500 gallons at full opening—only a lucky few escaped, and some of those were captured by seabirds attracted to the splash. A herring gull perched on the edge of the mouth as if on a birdbath until the whale slipped below the surface. The bubbles disappeared. The sea went calm.

Off the stern, a huge vertical spout rose, deep as a foghorn, then glints of stainless steel flashed off a slate-blue flank that arched above the water: a finback, large and fast—more streamlined than the humpbacks—passed like a bullet train, ending with a relatively tiny dorsal fin, the mammal so big that even the time it had taken to surface had seemed enormous.

Near the horizon, a humpback breached, twisting 360 degrees. A calf began to lob tail, moving its dark fluke in the air. Humpbacks are the splashy ones, playful, interactive. To Melville they were "the most gamesome and light-hearted of all the whales, making more gay foam and white water generally than any of them."[5] To at least one biologist, they were cute but exhausting, like three-year-olds or puppies. "They're always like, 'Look at me. Look at me.'"

And then there were the right whales, the first cetacean I had got to know, rare and brooding, plying the waters with an enormous scowl and a train wreck of cornified skin covered in whale lice, a callosity. They had already left for the more productive deep waters of the Bay of Fundy by the time we arrived on Stellwagen. These surly Goths of the North Atlantic had recently been dubbed the urban whale—they feed near Boston and raise their offspring off Jacksonville, Florida, one of the busiest ports in the country. One biologist has likened it to raising your kids on the interstate: there were so many enormous 800-foot ships in the region that these 40-foot whales were little more than possums to an eighteen-wheeler. Their populations, greatly reduced by overharvesting a hundred years ago, have been protected since the 1930s—but until recently, only against intentional take.

As Michael Moore of Woods Hole Oceanographic Institute, who has done more than his share of whale necropsies, likes to say, the United States is still one of the biggest whaling nations on Earth—but we do it through negligence, with ships and commercial fishing gear, rather than with harpoons. When caught in a net or line, humpbacks relax and let a team get to work, but right whales, Moore has discovered, need sedation. They're both subject to the same risks, but humpbacks have rebounded to more than ten thousand in the North Atlantic, with estimated growth rates of more than 3 percent a year. For a large-bodied, long-lived species, that was quite good, testament to the success of the moratorium on commercial whaling put into place in 1986.

As we floated over Stellwagen Bank, Boston nothing more than a callosity on the horizon, suddenly there was a spout of coral, then rust. Up ahead, a whale rose, arching her back and kicking with her flukes. Somebody called out, "Poop!" It was as big as our Zodiac, a plume of weak green tea. This cloud of unknowing descended several meters down the water column. I eased a plankton net through the plume and captured a bit in the cod end, stowing it in a cooler.

I had worried that, collecting and processing fetid fecal samples, I would get seasick aboard the *Nancy Foster,* which rolled, if not exactly like a well-greased pig, then like one in clover. But humpback feces came up roses compared to that of right whales. Roz Rolland, a senior scientist at the New England Aquarium, had used the right whale's stench to her advantage. She and her colleagues had been collecting right whale poop opportunistically for years, but Rolland wanted a lot of samples for her studies. So she employed Fargo, a 90-pound, six-year-old Rottweiler, to help her on the Bay of Fundy. Right whales live on copepods (tiny crustaceans); depending on the season, and the life stage of their prey, right whale feces is bright red or a deep muddy brown. No matter the time of year, the stuff was nasty, almost as bad as a right whale's breath. It's no wonder ancient mariners had feared that inhaling the whale's toxic breath could cause dizziness and fainting fits, possibly death; just a few droplets was said to raise a rash on human skin. Rolland said, "You get that poop on you, you have to throw the clothes away."

Fargo's nose was well-proven—one of his most spectacular feats had been locating right-whale scat at a distance of a nautical mile.

But he had one minor flaw: mal de mer. Before they headed out, Rolland, a veterinarian, would administer 25 milligrams of Benadryl to prevent nausea. It could have been worse: Fargo's former partner, a mixed breed rescued from the pound, had had a fear of whales, curling up in the bottom of the boat whenever one surfaced nearby.

Rolland has used the samples to check lipid levels, revealing the nutritional status of each whale; to test for protozoan parasites; to assay hormone levels, which reveal sexual maturity, pregnancy, and stress; and to measure biotoxins found in harmful algal blooms— paralytic shellfish poisoning may be curbing the whale's ability to recover from centuries of exploitation.

"They're assholes," said Pekarcik, as a whale-watching boat swooped in, stopping almost on top of a group of feeding whales. The boat was so crowded with tourists, many barebacked or in baseball caps, that it listed toward the whales. "These are the days you just want to kill yourself when you're a naturalist," Pekarcik said, "when all of Times Square seems to have packed on to the boat."

"It looks like a cattle drive," Winn said, "with all the boats following those whales."

Whale watching, of course, is big business: tourists spent more than $125 million dollars on tickets and travel to Stellwagen in 2008.[6] According to whale biologist Roger Payne, it is essential that such visitors "become awestruck by whales." Whale watchers, not scientists, are going to determine their fate.[7]

Here are a few things that endangered species have done for local communities. Manatees, in the Class of '67, attract hundreds of thousands of visitors to Florida each year, where they spend more than $23 million to see the sirenids in Blue and Homosassa Springs.[8] Reef-based tourism around the Florida Keys is almost entirely dependent on the dominant (and federally listed) staghorn and elkhorn corals; the industry employs more than forty-three thousand people whose annual wage income totals $1.2 billion.[9] Reefs supply more than half a billion people with food and work, buffering coastlines from waves and producing sand for the beaches—each hectare of reef generates up to $130,000 of services. The bad news: more than two hundred species—a third of all reef-building corals—are at risk (from bleaching and other diseases), and the buildup of CO_2 from the burning of fossil fuels is likely to change the entire chemistry of the seas.[10] Only amphibians appear to be in a tighter death spiral.

Americans spent more than $120 billion hunting, fishing, and wildlife watching in 2006. That's more than the Super Bowl. It's more than professional football. It's more than was spent on all spectator sports, amusement parks, casinos, bowling alleys, and ski slopes combined. Hard to believe, until you consider that more than 71 million Americans spent more than $45 billion dollars observing and photographing wildlife. They spent the money on food, lodging, and transport, on guides and fees to access public and private lands, on bird food, binoculars, spotting scopes, and backpacking equipment, on nature magazines and guidebooks. This passion for simply watching nature resulted in more than a million jobs.[11]

A common complaint is that wild areas reduce the tax base in a community—I heard it in Boiling Spring Lakes. I heard it in Florida. But the Departments of the Interior and Commerce and the Census Bureau have been gathering data since 1955. The most recent study showed that wildlife watching brought in almost $9 billion in tax revenues to state and local governments. And this doesn't even include other local services such as storm protection or the provision of fish and freshwater, or global ones like climate regulation.

The figures for bird-watchers alone are staggering: there are 48 million in the United States, compared to about 33 million anglers and hunters. Most birders just enjoy keeping an eye on their feeders and the birds that visit their backyards; but around 20 million travel each year to see birds, averaging about two weeks on the road.[12] That's a lot of birders, and a lot of cash. Just as cities compete for stadiums and factories, communities should vie for parks and charismatic fauna. Whooping cranes in Aransas and Necedah, bald eagles at Mason's Neck, and ivorybills—well, maybe, in Arkansas.

It's all about the bird: many bird-watchers will stay in basic lodging to add another species to their list. When my wife and I visited a popular birding destination in Vietnam, the mold was growing down the walls of our room, the shower was a hose with a drain in the bathroom, the AC a mildew express. Midges, mosquitoes, and geckoes were everywhere. The leech socks we wore were only mildly protective, but on our hikes we saw, to name a few, an orange-necked partridge (endangered), a white-shouldered ibis (critically endangered), and a wreathed hornbill—of least concern, but still, it's a hornbill.

Beyond feeding the leeches, bird-watchers can help local communities, bringing in money and encouraging environmental protection. As birding is rarely big business, it's less likely to degrade the local culture than a beach resort, but it can be disruptive, even lethal to birds. A close approach can chase birds from their nest, wasting energy when food is short and leaving their young vulnerable. Cagan Sekercioglu, an avian biologist at Stanford, has suggested tying biodiversity to income. A dollar a bird and up to $20 for a threatened species? Visitors could boost local support for nature reserves if bird-watching tours made contributions to local nature organizations when they traveled to less-developed countries. A company whose clients are typically rich, well educated, and committed could add a premium of about 4 to 10 percent for conservation to the cost of a trip.[13]

There's always the risk that visitors will outnumber—or outrace—their subjects. As dolphin tourism grew in Shark Bay, Australia, the number of dolphins declined. A single tour operator had no discernable effect; but once a second boat began operating, one in seven dolphins left the bay, calving rates declined, and areas with no tour boats showed an increase in these small cetaceans. So the Minister of the Environment revoked one of the licenses as a necessary sacrifice to keep the dolphins—and the tourists—in the bay.[14] Shark Bay, remote and small, was a pretty easy call. On Stellwagen, more than a dozen whale-watching companies have agreed to voluntary guidelines created to avoid whale strikes and to keep whale-watching vessels from pursuing, tormenting, or annoying them. Will that prove good enough for the whales?

Wiley, slight, with a shock of white hair, ice blue eyes, and a gray beard, was well known on the bank as research director of Stellwagen. In 2003, his less-conspicuous colleagues, traveling undercover but armed with GPS units, rangefinder binoculars, and digital compasses, found that noncompliance was common: boats traveled at high speeds to reach the whales, then raced back to port. Even around the whales, they used the vessel's maximum speed.[15] When Wiley asked one of the captains why he went so fast, the man said his passengers wanted to get back to land.

"But we gave you posters to explain why you have to go slow," Wiley reminded him.

"We took them down," said the captain. "People kept complaining that we were going too fast."

Wiley didn't start at Stellwagen. His first offshore assignment after graduating from the University of Massachusetts was as a marine-mammal observer—off Dumpsite 106. Back then, New York City and New Jersey shipped their sewage twelve miles offshore, dumping an average of eight million tons each year on the continental shelf. Bacterial levels rose. Heavy metals contaminated the seafloor. Were the dolphins and other whales in the area affected? There weren't many in the area, but Wiley had learned something critical to my analysis: the sludge sat above the thermocline, the border between the nutrient-rich bottom waters and the light-filled upper surface, exactly as I suspected whale poop did. Whale feces could enhance biological activity—but the millions of tons of concentrated and contaminated human sludge created an anoxic environment, a "dead zone" like the vast areas of the Gulf of Mexico around the mouth of the Mississippi River, where oxygen levels were so low, most fish and invertebrates couldn't survive. Shellfish beds were closed. Fisheries were closed. New York City finally stopped dumping its sewage there in 1992, but around the world, dead zones are still growing, caused in many cases by the runoff of excess fertilizer.

Away from these dead zones, the upper layer of many coastal systems becomes nitrogen-depleted as the growing season proceeds. In the spring, plankton bloom as temperatures warm, and a boundary layer is formed between the cold, nutrient-rich waters and the upper surface. Only in this upper layer—the euphotic zone—is there enough light for photosynthesis. There phytoplankton, the base of the marine food web, grow until they use up much of the nitrogen, iron, or other essential nutrients.

Meanwhile, the cold dense water nearer to the bottom remains rich in these nutrients. Here's where the whale pump comes into play: after this boundary layer has formed, many cetaceans actively feed at the bottom, rising to the surface to breathe—and poop.

At night, in two-hour shifts, we stood on the darkened bridge in isolation, in a bubble net of static, listening for a tag. It was like living in a TV set after the station had gone off the air—loud enough to hear a subtle ping at one-and-a-half miles, low enough so it didn't violate the Geneva Convention. "It's been eight minutes since we last heard a signal!" We started to worry that we would lose the tag. For the whales, it seemed, the days were for surfacing, the nights for diving.

The rest of the team mustered in the computer lab, where Colin Ware had been working on the data downloaded from an earlier tag. Director of the Data Visualization Research Lab at the University of New Hampshire, an expert in texture, movement, and 3D displays who designed ocean-flow models and taught a class on color, he turned on the LCD. Becky Woodward had spent the evening orienting the tag on the whale—the angle of the tag on the whale's back influenced how the tracks would appear on the screen—and then she had oriented the whale on the world, guided by shallow dives and fluke strokes.

The whales looked as if they were on an escalator shorn loose from its tracks, or in an Escher painting done with lavender and blue. Why those colors? "You want to use softer colors so you can see the shapes and shading," Ware explained. We watched the whale dive, hatch marks ticking off the depth. When it rose, the humpback's avatar bumped up against a green circle, Ware's representation of the surface. The whale made a large loop, slapped its tail, then swam in a tighter curl—a "feeding loop."

I had been following whales intermittently for more than a decade, almost always from boats or planes, occasionally from the shore. Watching Ware's tracks was a revelation. Almost everything I knew about whales had come from the time they spent at the surface, where they breathed, slept—and, yes, shat. Humpbacks had occasionally breached; right whales, as moody as they looked, had sometimes put on extraordinary sexual acts that could last for hours: surface-active groups featuring one female, several males, and an occasional sea snake, a nine-foot penis. But here was a day in a whale's life on a field of white.

Some whales followed a regular pattern: feed early in the morning (around seven to nine) and again in the afternoon, then rest for much of the remainder of the day. Other whales fed all the time, either by making bubble nets near the surface or by turning side rolls at the bottom. There were right- and left-flippered whales: some always made feeding rolls to the right, others to the left. You could see the handedness when they surfaced: righties had scuff marks along the right side of their bodies where they had hit the bottom in search of food. Some did feeding loops—one big loop near the surface followed by a smaller, tighter one where the whale captured prey.

Biologists had long thought that bubble-net feeding was cooperative—two whales would swim circles beneath a patch of forage fish, emitting a dense web of bubbles, then rise through it, feasting. But the first time the biologists watched two tagged whales feed together, they were surprised. One whale dove deep to school the sand lance, the second whale coming in only after the fish had been forced upward. "It's mooching off somebody's bubble cloud," Wiley said.

"Cooperative feeding is really stealing," Weinrich has concluded. "The second animal is not contributing anything at all.

One does all the work and the other exploits it." I thought of Mc-Intyre's metaphor about wolves—some were born hunters, others content to stay home with the kids; such pack-mates almost always shared.

Most associations are ephemeral, a day at best, but there are also some long-term ones. "For the past five years, we've had these two animals, Echo and Tectonic, who had never been seen together before," Weinrich continued. "Then they started traveling and feeding together. During that time, Echo has had two calves. Neither were Tectonic's. He seems purely random. Why associate if there's no relationship, no feeding, and he isn't even breeding?"

On the breeding grounds, the relationship is obvious: escorts are usually dominant males—the more time they spend with a female, the more likely they are to sire offspring. Weinrich has enjoyed destroying stereotypes. "Dolphins aren't all that nice," he has said. The first cases of infanticide in whales were observed in bottlenose dolphins: the majority of dead calves stranded in Scotland had multiple traumatic injuries, a sign that they had been killed by adults. Harbor porpoises are about the same size as dolphin calves: the adults killed them, too.[16] Female dolphins are covered in tooth marks during the breeding season, a sign of sexual coercion. Males are even more beat-up. In one population, four out of every five dolphins showed scars from scuffles. Almost all of them came from the males; in hundreds of hours of observation, females were seen to act aggressively only a handful of times.[17]

As we followed the avatar, we could see the whale's past as well its future, as if all its moves were predetermined.

"I think I see a breach in its future," Wiley predicted drily. The whale hit the surface, then angled back. Far off, in the distant white space of the future onscreen, there was a crash of bright red on the track, the end of the line—not a harpoon, but a dying tag.

Back in the ship's wet lab, my fears of seasickness proved unfounded. The humpback specimens smelled mostly of brine; there was the slightest bit of ash—some sand lance scales and bones—at the bottom

of the liter jar. I filtered the fecal samples and added reagents to measure the nitrogen in the plumes: the darker they turned, the higher
the concentration of ammonium. I ran the spec—ammonium levels
were through the roof for those that were deepest blue. Although the
ambient levels approached zero, the water from the fecal plume had
a concentration of more than 30 micromoles: the humpbacks were
releasing plumes of nitrogen more highly concentrated than the rich
bottom waters where they fed. Here was our first field evidence that
whales were fertilizing their gardens.[18]

Just as trees had become more than board feet or timber, whales
were far more than the number of barrels inscribed in a logbook or
the number of pounds of *kujira* or *hvalkjøtt* on the market. Whales
could increase primary productivity in the gulf, helping to sustain
fisheries and even, perhaps, fight climate change by pumping iron, a
limiting nutrient, to the surface of the southern oceans. Despite attempts to show that whales were our competitors—they eat our fish,
therefore they should be caught—it looked as if, in fact, more whales
meant greater productivity and more fish. Just increasing the standing stock of whales could help, their massive bodies sequestering
carbon after they died, like fluking forests in the seas.

Maybe it was better to watch whales than to eat them. But whales
were easy. What if we found out that returning species like rare mice
and wading birds were one of the best ways of protecting human
health?

14

Dutchess County, New York, about ten p.m. I had been told that the entrance to the Institute for Ecosystem Studies in Millbrook had a clear sign. But as I drove up and down the country road, my headlights shining through the tunnel of trees, I couldn't find it. There was the occasional flash of a firefly. The blushing crescent moon revealed a few stately homes.

Something jumped out in front of me at a twist in the road. I slammed on the brakes. A tawny young stag, a two-pointer, I think, bounded over a lapsed stonewall, its tail a white flag that fluttered in my eyes long after the creature disappeared.

The first thing I thought about—after my heart stopped pounding and I finally found the entrance—was ectoparasites: the animals that made a deer's skin crawl. (Let's forget for the moment about the internal worms, nematodes, and microbes.) Deer carry fleas. They carry mites. They carry lice that bite and lice that suck. They carry keds, parasitic flies that lose their wings when they find a warm, furred host and spend the rest of their lives on a single deer. The deer carry a lot of ticks: *Amblyomma americanum*, the lonestar tick; *Dermocentor albipictus*, the winter deer tick; and *Ixodes scapularis*, the black-legged tick, so closely associated with that stag and its kin that it was known as the deer tick. I was here to see it.

I made my way down the dirt roads of the institute to Knapp House, my lodging for the night. With its thick green shag rug, dark paneling, and large deck, the house felt as if it had fallen out of the polluted skies of the 1970s. I heard bullfrogs calling for their *jug-of-rum,* and a single spring peeper. A dehumidifier roared through the house, struggling to keep mildew at bay, but the smell seeped through the rooms.

Back in the 1970s, there was an outbreak of skin lesions and swollen joints among the children of the quaint, coastal town of Lyme, Connecticut. Allen Steere, a fellow in rheumatology at Yale who had worked at the Centers for Disease Control and Prevention, diagnosed the illness as pediatric arthritis and dubbed the condition Lyme arthritis. The following year more cases were reported, this time from the Naval Medical Hospital in nearby Groton. These were wealthy areas; early on, Lyme disease was seen as an affliction of the well-to-do, of those who could afford "forested isolation."[1]

Early on, misdiagnosis and overtreatment were common. Steere identified the pathogen as a virus, the vector as a tick. The children of Lyme got sicker. Some patients received steroid shots, which reduced inflammation but could increase infection. One woman, operated on for a brain tumor, was discovered to have Lyme; she was given antibiotics and recovered. Another patient, treated for Lyme, was found to have a tumor instead.[2] The cause of the disease was, as Willy Burgdorfer discovered in 1982, a spirochete bacterium, not a virus. Found from Maine to California, Lyme costs about $1 billion a year in medical care, lost productivity—even legal fees. The long-term effects of *Borrelia* infection, or chronic Lyme disease, include memory loss, joint pain, irregular heartbeat, and chronic fatigue. Depression, job loss, and divorce have followed infections, though the existence of a chronic disease long after the bacteria have been eradicated is hotly debated in medical journals. The standard treatment is a two- or three-week course of the antibiotic doxycycline.

The blame for the disease is often laid squarely on the shoulders—and the ears, hind legs, and base of the bouncing white flag—of the white-tailed deer. The traditional narrative is that the restoration of *Odocoileus virginianus,* and the forest that supported it, prompted the emergence of Lyme disease in the Northeast. It reshaped our view of the wild and its leggy inhabitants. The vector was after all, a *deer* tick. Rick Ostfeld, the disease ecologist I had come to see, has a different angle on this emergence.

In the morning, I met Kelly Oggenfuss and Erica Dolven-Kolle in the institute's lab. Before they would take me into the woods, I had to get into a white jumpsuit. I was glad to have it. When some people walk through the woods, their fears might turn to bears or snakes. Others are hounded by mosquitoes or horse flies. I've always been a tick magnet, emerging from the woods with several deer or dog ticks, heads deep under the skin. They're fairly inconspicuous at first, paper-thin parasites until they bum a meal. Fully engorged, they increase their body size by an order of magnitude, shape-shifting into a potato with claws.

The jumpsuit hung a bit loosely, but at least it was relatively new—Oggenfuss and Dolven-Kolle looked like forgotten cosmonauts, their suits held together by broad swaths of duct tape. I refrained from asking how they would discover if their suits had holes.

"Hey, tick crew," two young women yelled as we crossed the parking lot. "We love ticks!"

We piled into a bright blue GeoTracker, a toy of an off-road vehicle that got us into the hills on the eastern edge of the institute's grounds, passing a group of Girl Scouts along the way. I followed Oggenfuss through the woods, the overgrown grounds of a former estate. Working the traplines, she grabbed a white-footed mouse with her gloved hand, catching it in a yawn of fear. She blew in its fine light brown hair in search of nymphs, a field of wheat in the breeze. This particular mouse was clean: no ticks. When she let it go, it disappeared beneath a log. A badly infected mouse, Oggenfuss told me, looks like a poppy-seed bagel.

Mice populations vary with mast, the amount of acorns on the forest floor. Though we couldn't see it, the woods were alive with mice, about 25 per acre. They convert acorns to flight, in the form of owls and hawks; to a midnight scamper by weasels and raccoons; or to a slither in the grass—snakes. The importance of mast has led one researcher to hypothesize that the end of the passenger pigeon might have played a key role in the rise of Lyme disease: fewer birds meant more mast, more mice, more of the spirochete *Borrelia burgdorferi.*[3]

Marked with a red flag sticking up from a PVC pipe, each station had two small aluminum Sherman traps, baited with oats—just a teaspoon's worth of grain to avoid changing the forest dynamics. As Ostfeld, the principal investigator on the project, would later make clear, he didn't want to subsidize the diet of the rodents in the area and elevate their numbers. We continued up the line, crossing over old trees fallen across the forest floor like pick-up sticks. A gypsy moth infestation had devastated the area several years ago—and it turns out the mice play a role in this cycle, too. White-footed mice are principal predators of the gypsy moth and can help keep this invader in check, but their populations fluctuate with acorn harvests. A bad year for mast is a bad year for mice and a good one for gypsy moths. This can push a forest into a downward spiral as moths destroy more trees.[4]

The next trap had a chipmunk. Oggenfuss snapped it out of the thin metal box as if she were swinging an ax. The terrified rodent landed in a Ziploc bag, reinforced with—what else?—duct tape. Oggenfuss squeezed it into the corner of the bag like toothpaste from a near-empty tube. Keeping its nose in the corner, she grabbed the chipmunk by the scruff of the neck and pulled it out. She noted the sex, whether it was nursing—lactating females are released—and weighed it. Those with ticks were put back in the trap, those without released to let out a screech of protest when they were out of reach, about five yards away. One stopped at Oggenfuss's foot. "It's not a tree," she said, "I wouldn't go up there if I were you." It disappeared into the leaf litter.

The black-legged tick has a two-year life cycle. Females deposit eggs under the leaf litter in the spring; the minute larvae, pink as newborn mice, were hatching when I visited in June. At each of the tick's three stages—larvae, nymph, and adult—it requires a blood meal to molt to the next stage or to reproduce. The larvae emerge from the litter questing, looking for their first meal, front legs held high, hoping to hitch a ride on a small mammal: a white-footed mouse, chipmunk, or squirrel. Six-legged at this stage, the larvae are so small they can quest on the head of a pin or the tip of a blade of grass. Those that succeed molt to eight-legged nymphs about the size of a poppy seed, almost invisible—but, to some, nature's true marauder. Before molting and becoming adult, nymphs feed on mammals or birds for their second blood meal. Around here they mostly feed on mice. If one captures a human, the tiny nymph just might escape notice long enough to transmit disease. Before overwintering and laying eggs in the spring, the adults mate on deer and other large mammals. Ticks love warm dark places—on humans, the hair above the nape of the neck or underarms if they can get there.

As the feeding tick swells and its body temperature increases, the dormant spirochetes begin to rouse. The proteins in their outer membrane change; they become infectious, multiply, and move from the gut to the body cavity to the hemolymph, invading the salivary glands. Chemicals in the tick's saliva suppress the immune response of nearby skin cells, aiding the transfer of *Borrelia* to the bloodstream of their new host. That it can take a day or two for the spirochetes to reach the tick's saliva is one of the reasons that adult ticks rarely pass on Lyme—a person is unlikely to miss a blood-filled lentil-size parasite for that long.

On reaching the blood of its host, the bacterium first infects the skin. In humans, erythema migrans, a red rash with a central clearing or a bull's-eye, typically erupts near the bite, one day to a month after infection. Later the spirochete disseminates to other epidermal regions, and, eventually, to the joints, the heart, and the nervous system. Untreated, Lyme disease can cause paralysis and a type of palsy, making half the face go slack. It can also cause arthritis and nerve damage, swollen glands, buzzing in the ears, night sweats, bad hangovers—even seizures, memory loss and confusion that at its most extreme can resemble dementia, irregular periods, and milk production in women who aren't pregnant. Symptoms show up in people, horses, and dogs.

Oggenfuss started working at Millbrook the summer after she graduated from Rutgers in 1999. Did she ever have Lyme disease? "Only once since I've been here." Her symptoms included multiple rashes around her lymph nodes, muscle ache, headache, and fever. Antibiotics had rid her of *Borrelia,* though she had been extremely fatigued throughout the entire treatment. How long was she out? "I may have been out of work one day to see a doctor."

We lined up the Sherman traps, like cheap metal caskets, in the back of the Tracker. A vulture circled above, with wings like steely fingers. But this wasn't a death march. The mice and chipmunks would spend a few days in the rearing facility, where they were kept in small wire cages suspended over water pans. The captives would have air-conditioning, free food, and no predators until the last of their ticks finished its meal and fell off.

Blood meals aren't fast food. At the rearing facility, it takes three days for all of the larval ticks to engorge and drop off their hosts. Drowning is never pleasant, and the arachnids suffer the indignity of drowning in their host's feces (what the researchers call "poop water"). The dead larvae are brought to a desiccating chamber: picture a take-out salad bowl. Then each individual is moved to a vial where plaster of Paris sucks out the last of the moisture. The hosts are returned to the forest, to continue their role in the ecosystem—and in Rick Ostfeld's long-term studies.

I found Ostfeld in his tidy office. He wore wire-rimmed glasses, a button-down shirt, cargo shorts, and sneakers; his clothes looked pressed and he looked pumped, as if he had just been to the gym. Beneath a cliff of books—vertical slabs of mammals, forests, fungi, and bacteria—he told me the tale of a black-legged tick, a white-footed mouse, and an emerging infectious disease.

"It's a grind-it-out, long-term monitoring project," Ostfeld laughed, "all to get one landscape-level data point per year." The pace of science can be excruciatingly slow, like a novelist writing a page a year or a poet a stanza, each word a day spent in the field counting ticks, running trap lines. It's not for the faint of attention.

That white-tailed deer I had seen by the edge of the institute—was it really something to worry about? "The role has been grossly overstated," Ostfeld told me. "Adult ticks mate while on deer, but it doesn't matter how many adults you have. Without small mammals such as mice to provide a blood meal, the larvae can't grow up to be nymphs or to transmit the disease. The number of nymphs is dependent on the number of mice, not the number of larvae."

When the earliest cases of Lyme disease emerged, some researchers made a connection between the high rates of childhood arthritis—mostly swollen knees—and tick-borne diseases in Europe, which had similar symptoms. Biologists on Long Island examined the carcasses of deer brought in by hunters and thought they had found a new species of tick, which they named *Ixodes dammini,* the deer tick.

Those early studies were mostly done on islands with reduced faunal diversity. Even the tick itself was misidentified: it was actually a northern variant of a common southern species, *Ixodes scapularis.* But the name, and the stigma, stuck. "On islands, there was no other game in town for adult ticks than deer, and there was no other game in town for immature ticks than white-footed mice. So the dogma arose, and persists to this day, that deer are indispensable for tick populations and white-footed mice infect them. That's the dogma. At best it's too simple, and at worst it's just flat-out wrong. I feel fairly alone in this, but I think the evidence for that scenario is inadequate."

An ecologist stranded among epidemiologists. I thought of a marsh bird caught up in a Category Four hurricane—surviving the storm, he alights on a new island, only to the find that the natives want nothing to do with him. If he's fit enough, if he can survive the selection pressure from established species, he just might thrive in his new niche. He'll still need someone to . . . um . . . talk to, of course—Ostfeld's closest colleague is his wife—and a place to disseminate his ideas (this is a rather intellectual bird). The new journal *EcoHealth* provides one such ecosystem, distributing papers that integrate human, wildlife, and ecosystem health. New departments of eco-epidemiology and conservation medicine provide habitats where ecologists, conservationists, epidemiologists, and physicians can work together as mutualists, do some research, publish papers, and maybe, just maybe, reduce disease and extinction risk.

Although it was first isolated in 1982, *Borrelia burgdorferi* has probably been in North America for millennia. Genetic analysis suggests that it could have been on the continent for a million years.[5] "Presumably, Native Americans were occasionally getting sick," Ostfeld told me, "but they had bigger health problems to worry about, so it wasn't recorded." The same was probably true of the early European colonials in North America—they got Lyme, but a little arthritis was nothing compared to starvation, trauma, and other diseases that were a lot more obvious. During the seventeenth through mid-nineteenth centuries, New England, an area that was once 90 percent forested, became 90 percent deforested.

"There was enough wholesale forest conversion and destruction that ticks and some of their hosts were virtually wiped out over most of New York, New England, and the Mid-Atlantic areas," Ostfeld continued. "The evidence suggests that ticks and Lyme disease survived in refugia along the edges of Long Island Sound. Of course, all of these farms were abandoned wholesale in the eighteenth and early nineteenth centuries. The opening of the Erie Canal, when the shipment of grain to the Northeast became cheaper, sent farmers off to the Ohio Valley and Mid-Atlantic areas. New England was a terrible place to farm, you know. The colonials were starving every few years. So reforestation took place, and now we're back to about 70 percent forest cover.

"Mice probably bounced back relatively quickly, as did the other hosts of the ticks. The thinking is that during the mid- to late twentieth century infected ticks spread from these refugia around Long Island up into central New England, New York, and down to the Mid-Atlantic. The timing of all this is a little bit mysterious. There are a lot of people who say, well, deer expanded very quickly from near extirpation and that's responsible for the subsequent spread of Lyme. But there's a multidecadal time lag between the time the deer expanded and the time that the ticks expanded. I don't think the expansion of the deer tells us very much."

There's some experimental support for Ostfeld's challenge to the deer dogma. Research in the Italian Alps showed that the number of nymphs increased in areas kept free of deer. The ticks in deer exclosures fed almost exclusively on rodents, which had high rates of tick-borne encephalitis, creating hotspots for the disease. The presence of deer actually reduced the transmission of the pathogen, at least on a small scale. The exclosures where tick amplification occurred were smaller than 2.5 hectares, about the size of a typical forest fragment in the suburbs.[6]

Deer aren't even very good hosts for *Borrelia*. A few drops of deer blood added to a culture of it will lyse the cells and kill all the bacteria. To this spirochete and other parasites, all of us—deer, humans, mice, ticks—are just ports in a storm, temporary islands for the effective transmission of genes. Lots of the familiar animals of the eastern United States—squirrels, deer, voles, raccoons, opossums, and skunks—are poor reservoirs for *B. burgdorferi*. Their bloodstream isn't a great ecosystem for the bacteria; the immune systems attack it, or it just barely holds on. Opossums are tick magnets, often carrying more than 250 of them, but they're an epidemiological dead end, a marsupial eraser; they feed ticks, but don't infect them with the Lyme spirochete. We're also dead-end hosts—if a tick bites an infected human, it won't pass on the disease.

In the other corner is the white-footed mouse, weighing in at less than an ounce, perhaps the ideal habitat for *Borrelia burgdorferi*. The bacterium doesn't seem to impact the mouse, though, as Joe Piesman of the Centers for Disease Control and Prevention said, "Whether it decreases mouse happiness, we don't really know." More than half of the ticks that feed on *Peromyscus* acquire *B. burgdorferi*. Shrews and chipmunks also make pretty good hosts for the pathogen.

"There's competition for ticks among the host mammals. We've analyzed tens of thousands of mice and chipmunks in the past ten or twelve years. We actually have quite good evidence that other hosts in the system deflect tick meals away from mice. The more chipmunks you have out there, the lower the per-mouse tick burden. You got all these ticks out in the environment—and those that bite white-footed mice are more likely to survive and feed again later on us or something else. They are much more likely to acquire the pathogen." Ostfeld and colleagues call this the dilution effect: increased species richness reduces the presence of zoonotic disease. The trouble is that as we fragment the landscape, we're creating our own exclosures, landscapes free of many of the larger mammals and predators that can dilute disease.

Critics remain skeptical. One epidemiologist is emphatic that Ostfeld and his colleagues should be challenged on every level, at every chance. Durland Fish, director of the Center for EcoEpidemiology at Yale, confronted the dilution effect with his own coinage: the "pollution effect." More species mean more reservoirs for disease. He cited the recent emergence of the tick-borne Powassan virus in Ontario as a possible result of the return of fishers—a large marten—and other mustelids to North America. "If we got rid of all species," he said to me at a conference, "we wouldn't have zoonotic diseases."

Of course, we'd also lose all the goods and services provided by animals. We'd be vegans—alone on a planet of soy and the other plants we could pollinate ourselves.

"The tendency within the biomedical community is to try to find the minimal number of players necessary to explain the health effect," Ostfeld told me, "and then to stop." You identify the pathogen, the vector, and the hosts—in this case deer and mice. "I think that approach has failed them. If we say that some laboratory study found that you can infect skunks with Powassan virus, and they infect the ticks—boom! therefore, mustelids are the most common reservoir for Powassan—and stop there, I think that does a huge disservice to understanding ecological systems. You can't stop there, because you have not pursued the important things. You have to raise the complexity. You have no choice. It's not as daunting as you might think, but it requires more holistic studies. Biomedical experts and medical entomologists—they don't want to do biodiversity studies. They're not trained to do it. They're not interested in it. But I would argue that you have to do it."

We have to do it, not only to understand these illnesses but also to defend the innocent. As the most commonly diagnosed vector-borne disease in the country, Lyme has changed our perception of nature and the degree to which people find comfort or solace in the woods. "Forests have become a threatening part of the landscape," Ostfeld said. "People have stopped gardening. How many people no longer hike? How many people no longer let their kids go running through the woods or don't hunt anymore? Or do so with trepidation and fear? I bet the answer is an enormous number."

For a moment, the trees outside Ostfeld's office window seemed to be watching us, an iris of maple and birch, an immense emerald eye. Whether you saw benevolence, as this eye twinkled in the breeze, or a piercing cold gleam was in part a matter of upbringing and epidemiology. Was it a place to regain your health or a sanctuary for disease?

Throughout human history, changes in culture have brought about shifts in disease ecology. Some anthropologists have proposed three major epidemiological transitions, which follow the course of our cultural history.[7] The first occurred during the shift from hunting to agriculture, around eleven thousand years ago. At the end of an ice age when the climate warmed dramatically, then stabilized, after more than a hundred thousand years of restless hunting and gathering, we began to settle down.

It may seem counterintuitive, but the healthiest time in our evolutionary history was probably when we were on the move. When people dispersed from Africa, they left behind the humid, densely populated, and parasitic tropics. On the African savannah, humans were chronically infected with hookworm and roundworm, typhoid and dysentery. But, as they moved north to the cooler, drier climates of the Middle East, Asia, and Europe, these small bands of hunters and gatherers didn't have the numbers to keep acute infections running through the population.[8]

The new agriculturists weren't as well fed as their ancestors; they were shorter, and many suffered from anemia and other nutrition-related ailments.[9] Along with farm tools, the Fertile Crescent has yielded the earliest evidence of war: defensive walls, maces, and skeletons pierced by arrowheads found in Iraq, presumably to defend stockpiled food, date from this time. In shifting to agriculture, humans began living with animals rather than just chasing them. We benefited from this new reliable source of protein, but there were costs: an increase in microbial traffic—domestic animals provided a bridge for infectious agents to move into human villages and towns—and the evolution of more virulent forms of existing human pathogens in now sedentary populations. Diphtheria, influenza A, measles, mumps, pertussis, rotavirus, smallpox, and tuberculosis: these are a few of the highly lethal diseases we got from domestic animals as we became agrarian.[10] Schistosomiasis—a chronic disease causing liver and intestinal damage—became endemic among peasants working in the irrigated agricultural fields of Egypt and Sumeria.

The emergence of a new disease occurs in stages. An animal pathogen (be it viral, bacterial, fungal, protozoan, or metazoan) is introduced into a new host, often from another species. In the early stages, the pathogen has to be transmitted to humans from an animal host. Rabies and West Nile virus, for example, aren't passed between humans. During intermediary stages, diseases can be transmitted between people, with periodic outbreaks from wildlife. Ebola and Marburg are in this stage. The final stage is when the disease becomes exclusive to humans: faliciparum malaria, smallpox, or syphilis.[11]

Infection is intrinsic to life on Earth, but the great majority of microorganisms never come near humans. Of those that do infect us, most are benign, some beneficial. Even infectious agents intend us no harm. They're just trying to make a living like everyone else: maintain energy, absorb nutrients, reproduce. Pathogens used to spread gradually on land: long-distance travel was slow and human settlements scattered. A sick person traveling by foot was bound to be left behind, never to reach his destination; civilizations and their communities could remain fairly isolated. On the water, they picked up speed: with favorable winds, a sailing vessel could travel a hundred miles in a day on the Mediterranean Sea. A sick passenger could quickly contaminate the others onboard and a pathogen spread from Jerusalem to Athens or Rome; the sea and its shoreline cities were essentially one disease pool.

A second epidemiological transition occurred during the Industrial Revolution. In pre-industrial societies, death and children were ubiquitous. Childbirth was often lethal, and childhood diseases swept through entire populations. Children helped raise their siblings. Once grown, they raised their own—and rarely lived much beyond the day their youngest left home. As life expectancy and the age of first reproduction rose in the nineteenth and twentieth centuries, chronic diseases emerged as the greatest health challenge. Water and air pollution were linked to higher rates of cancer, allergies, birth defects, and impeded mental development. Perhaps most importantly for human health, Robert Koch isolated the cholera bacillus in 1884. The assault on infectious disease had begun. It culminated in a fight against one of the most lethal viruses in human history, the first premeditated attempt to drive a species into extinction.

Smallpox had emerged about ten thousand years ago in North Africa, probably making the species leap from cows to humans— cowpox is its closest relative. The virus had wiped out villages. It had killed princes and peasants. It had helped bring down empires: as many as seven million people had died of it during the decline of Rome. Doctors could recognize it with their eyes closed, Richard Preston wrote in his introduction to *Smallpox: The Death of a Disease*, simply by running their fingers over the pustules, as hard as buck-shot embedded in the skin. When the blisters erupted, a patient was reduced to a painful, dripping mass of pus. A smallpox victim gave off a stench, sickly and sweet, so strong it could fill a ward. The whites of the eyes turned blood red. Hemorrhagic smallpox, about 5 percent of cases, was invariably fatal, the victims hemorrhaging from the mouth, nose, and intestines.[12]

In 1967, the World Health Organization began a global eradication program, a year in which smallpox was reported in 42 countries, killing two million people. Less than a decade later, the deadly *Variola major* made its last stand on Bhola Island in the Bay of Bengal: a two-year-old Bangledeshi girl was treated and survived in 1975. *Variola minor,* or endemic smallpox, was last seen in a hospital cook in Somalia in 1977. Both forms of the virus are now extinct in the wild.

Many thought that eradication of other infectious diseases was only a matter of time and effort. Worldwide campaigns would systematically isolate and cure each one, from measles to yellow fever. Or so it seemed at the time. But then new drug-resistant strains of familiar diseases, such as tuberculosis and malaria, began to re-emerge. Perhaps the strongest counterattack to these efforts—and the biggest blow to the optimism of 1976—came from two single strands of viral RNA.

On June 5, 1981, the Centers for Disease Control reported in its weekly bulletin that five severe pneumonia cases had been reported in three Los Angeles hospitals. Two things made these cases unusual: all of the patients were young homosexual men, and they all were infected by *Pneumocystis carinii,* a widespread protozoan known to cause serious illness in newborns, but rarely showing symptoms among adults unless their immune systems were compromised.

Soon the infections were not isolated to LA. Severe cases of pneumonia had been reported in New York, giving rise to rumors that a new malignant disease was hitting the gay community. By the end of the year, 408 cases had been reported to WHO, 5,077 in 1983,

12,174 in 1984. The real situation was much worse: Africa was severely affected by the pathogen, but public health authorities would deny the fact for years. HIV-1 took the world by storm. That the source of the disease is likely an endangered species—the common chimpanzee, more common when the disease made its leap from *Pan* to *Homo*—is ironic. Probably passed through the consumption of bush meat, the virus has been amplified by the movement of people from rural to urban areas. Now, loggers are bringing AIDS back into the forest they're cutting down, spreading it to women who have had to turn to prostitution in regions that were once isolated from many communicable diseases.

The consumption of wild animal meat—bush meat—is one of the primary highways for microbial traffic. More than a million tons of wild animals—mostly duikers, pigs, primates, and rodents—are harvested in the Congo Basin each year, the equivalent of about four million head of cattle. (Americans consume far more meat per capita, though much of that comes from feedlots, not forests.) Logging roads have made market hunting profitable: it is now a primary driver of many economies in the region, providing up to 80 percent of the dietary protein and fat in rural communities.[13] The diseases directly released from these hunts—HIV/AIDS and Ebola are two—have caused millions of deaths and hundreds of billions of dollars of economic damage globally.

The deathblow to many established infectious diseases never came. Malaria and tuberculosis continued to kill millions. There are now between 50 and 100 million debilitating cases of dengue fever a year; about 5 percent of those who get dengue hemorrhagic fever, the most severe form of the disease, die. In addition to HIV/AIDS, other new afflictions have emerged: rotavirus, cryptosporidiosis, Legionnaire's disease, Ebola, hepatitis C, hantavirus, and toxic shock syndrome. The rise has prompted some anthropologists and epidemiologists to propose that we are now converging on a new global disease ecology: a third great epidemiologic transition. Most emerging infectious diseases are zoonotic, many residing in wildlife reservoirs.

Deforestation, irrigation, industrial agriculture, and other forms of ecological disruption have helped release these diseases into human populations, creating novel ecosystems where they can persist. Our closest relatives are not the greatest source of disease: ungulates, carnivores, and rodents have been the most common reservoirs, followed by primates, birds, bats, and marine mammals.[14] Many of these new pathogens are RNA viruses, adept at making species leaps. They have high mutation rates, enabling them to adapt to a new environment: us.

Perhaps the greatest risk of outbreaks occurs in markets like the one in Guangzhou—the presumed epicenter of severe acute respiratory syndrome, or SARS—which trades in masked palm civets, ferret badgers, barking deer, wild boars, hedgehogs, foxes, squirrels, bamboo rats, gerbils, snakes, and endangered leopard cats, along with domestic cats, dogs, and rabbits, allowing pathogens to jump between species that wouldn't normally be together in the wild.[15] It's not just an Asian problem. More than twenty-five thousand seizures of wild meat, including chimpanzee and gorilla body parts, were made in British airports in 2005.[16] Once marginal microbes spread and amplify when they reach urban areas, the traditional gateway for infections. Cities are monocultures, providing a bipedal habitat for new infections to become established, then spread out to other regions—more than half of all humans live in a city. A thousand years ago, when a disease emerged on the shores of the Mediterranean, everyone from Jerusalem to Gibraltar was at risk. The shoreline of that disease pool now stretches around the world.

In August 1999, six people from northern Queens were admitted to Flushing Hospital with high fevers and headaches. Early screens for the familiar bacterial infections came back negative. Three of the older patients died. Fifty-nine people were hospitalized in New York City that summer; the death toll rose to seven.

Duane Gubler at the CDC in Fort Collins, Colorado, examined the blood and cerebrospinal fluid of the victims and brain tissues from the deceased and identified a new strain of Saint Louis encephalitis, a viral disease spread by mosquitoes in North America.[17] Around the time of those earliest cases, there had been reports of die-offs of hundreds of crows and other birds throughout the city. This wasn't typical for that form of encephalitis. By the end of September, DNA sequencing of the virus came up with an exact match: a strain of West Nile virus that had caused an epidemic in Israel the year before. No one is sure how the disease got here, though it's possible that an infected bird was illegally imported into the country and then got bit by a local mosquito, which flew off with the disease.

The global spread of pathogens—West Nile virus, yellow fever, schistosomiasis, and malaria, to name a few—and their vectors offers the strongest link between the decline of native biodiversity and human health.[18] Invasive species have been a scourge of native ecosystems and human populations everywhere: as the world becomes homogenized, exotics disperse, native species decline, and infectious diseases become ubiquitous. The crew of the British whaling ship *Wellington* has often been blamed for introducing the first *Culex* mosquitoes to Hawaii, in drinking water from Mexico in 1826. The subsequent epidemics of avian malaria and poxviruses that broke out among native birds may have been initiated with this ship. Of the 136 bird species on Hawaii before humans arrived, 101 have gone extinct; many of the remaining species are likely to follow, threatened by invasive species, low population sizes, and disease.[19]

In the United States, one of the primary vectors of West Nile is *Culex pipiens,* the northern house mosquito, though many mosquito species can carry it. A true urbanite, *C. pipiens* breeds in tin cans, birdbaths, ditches, waste lagoons, and septic seepage, requiring little, if any, forest cover. It loves to feed on city birds: American robins and introduced house sparrows and starlings. Not until the migrating birds leave for warmer climes does *Culex* begin to look for other meals in earnest. Left with no avian blood meals, they zero in on another common vertebrate: humans.[20]

Around the time that West Nile showed up in Queens, severe epidemics of the disease occurred in Romania and Russia. Thousands were infected; about one in twenty had severe neurological disease. Since the earliest detection in the United States, more than sixteen thousand cases have been reported to the CDC. Unlike Lyme disease, which had been seen as a disease of the rich, West Nile hit low-income urban areas, with afflictions such as paralysis, stupor, and disorientation. More than seven hundred people have died.[21]

The disease was new to North America, but it had been isolated in the blood of a 37-year-old woman in Uganda in 1937, one of the earliest-known instances of a mosquito-borne virus. At first, the disease attracted little notice; many of those with infections had only a mild fever, headaches, or rashes—or no symptoms at all. The later bird die-offs, neurological disorders, and fatal infections in humans were, as one researcher wrote, "totally unexpected."[22] Something had changed in the virulence of the pathogen.

Humans carry the virus, but in general, we don't have a high enough titer, or concentration, of it to transmit the disease. Birds, on the other hand, are excellent hosts. Or at least some birds are. Is your backyard full of robins? Maybe a few sparrows or crows? These perching birds are highly competent reservoirs for the African virus.

Or perhaps there is a well-balanced suite of birds in your community—woodpeckers, wading birds, raptors? Your risk of catching the disease could go way down. In Louisiana, Vanessa Ezenwa and colleagues showed that infection rates of *Culex* mosquitoes declined with an increase of bird species richness, a measure of diversity. Waders, egrets, and herons are dead-end hosts—when a mosquito bites one of these birds, the disease stops there. With a smaller percentage of mosquitoes carrying the virus, the number of human cases drops.[23]

It is another example of the dilution effect. The point isn't to get rid of the robins but to restore a healthy community of birds, which could help fight any number of diseases. Even that turkey vulture I had seen riding the thermals over the institute as we packed the mice and the chipmunks away was a disease-control mechanism. Vultures metabolize the biotoxins in decaying flesh, checking the flow of infectious diseases such as botulism and anthrax transmitted via carcasses.

The virus itself might make bird-restoration efforts difficult. It continues to spread through the United States, though four out of every five people infected show no symptoms at all. Human deaths from West Nile may be rare but as the epidemic grew, there were steep declines in several avian species, including American crows, robins, chickadees, eastern bluebirds, and tufted titmice.[24] These links between diversity and disease risk may account for its distribution. When researchers looked at feathered diversity in the eastern United States with and without positive tests of human West Nile, the counties with the greatest species richness had fewer cases of the disease.[25] The endangered red-cockaded woodpecker and other picoides are dead-end hosts for West Nile—perhaps the folks at Boiling Spring Lakes would root for more clusters, if they knew the birds protected them from an infectious disease.

On May 14, 1993, Merrill Bahe was on his way to his girlfriend's funeral in Gallup, New Mexico, an area close to Arizona, Colorado, and Utah known as Four Corners. She had died a few days earlier of sudden respiratory illness. Though he was a long-distance runner, the young Navajo found himself gasping for air. He soon had a fever and a debilitating headache. His family called for an ambulance, but resuscitation efforts failed; Bahe died before he reached the emergency room in Gallup. The internist at the Indian Health Service, Bruce Tempest, struck by Bahe's youth and unusual death, called the state medical examiner. Tempest soon discovered that there were others in the region who showed a similar progression, including the girlfriend's brother, who died a few weeks later. Young adults were the usual victims, hit first with fever and muscle aches. Leaks in the capillary network surrounding the lungs restricted the ability to absorb oxygen; the patients gasped, starving for oxygen as fluids pooled up in the lungs. The heart slowed. Like Bahe and his friends, most victims died.[26]

Why were healthy youths succumbing to respiratory distress? In her book, *The Coming Plague,* science writer Laurie Garrett recounts that more than a hundred scientists, physicians, and animal trappers arrived in Four Corners. Tribal medicine men and a local ecologist provided some clues: the piñon nut harvest was large that spring, followed by a tenfold increase in rodents. When scientists trapped the area, they caught house mice, chipmunks, and several other species—but more than half the traps had brown, big-eared *Peromyscus maniculatus,* deer mice.

In the blood of the victims, the Special Pathogens Branch of the CDC soon discovered antibodies for a family of viruses common in rodents. Hantaviruses have been around for 30 million years, as long as there have been mice and rats. There are lots of strains: the Muroidea, which includes mice, rats, gerbils, and their relatives, is the largest mammal superfamily. The 1,300 or so species are found in just about every habitat on every continent except for Antarctica. Many of them get along quite well with *Homo sapiens.* Rodents are important reservoirs for many pathogens, including the agents of plague, monkeypox, rickettsial pox, Lyme disease, Rocky Mountain spotted fever, several types of encephalitis, toxoplasmosis, leishmaniasis, Lassa fever, and Chagas disease. There are at least nineteen viruses known to be associated with the house mouse alone.

Unlike West Nile and Lyme, which tend to hit cities and their suburban edges, hantaviruses are prevalent among agricultural workers. One in every three of the deer mice trapped in Four Corners carried the virus. Sin Nombre, as it came to be called after local residents, mindful of what had happened with Lyme, protested the name Four Corners virus, enters humans through the respiratory system. There's no need for a mosquito or tick to inject the pathogen: we inhale it from dried mouse feces. The disease kills about five hundred Americans each year.

One kidney expert soon wondered if the disease known as trench nephritis, which had mysteriously killed hundreds of entrenched soldiers during World War I and the American Civil War, was actually a hantavirus. By 1995, 33 different hantaviruses had been discovered in as many rodent species. More are added each year. Given the vast diversity of rodents—there are more than 2,200 described species—and their population booms and busts, shouldn't there be more disease where there's more species richness?

But disease transmission doesn't work that way. Unlike many wild animals, rodents are often found in our gardens, yards, storage sheds, even in homes. (Here I would like to thank our pet cats, who are housebound, for the service of clearing our cabinets, basements, and attics of many of the rodents that attempt to overwinter in our home.) According to Jamie Mills at the CDC, thirteen of the rodent-borne viruses that cause hemorrhagic fevers have been identified in disturbed habitats with low vertebrate diversity and few predators or competitors. Almost all the adult rats tested in Baltimore carried hantavirus, as did many of the house mice.[27] When a native tropical forest in Bolivia was cut, burned, and converted to sugarcane, only a few opportunistic rodent species survived. The clearing of forests on the edge of the cerrado in Brazil left tiny islands of native vegetation surrounded by huge plantations of soybean, sunflower, and sugar cane. Hantavirus is now endemic to both areas. Argentine hemorrhagic fever occurs in agricultural areas where its host, known locally as the *ratón maicero* or corn mouse, thrives on the weedy vegetation along field edges and roadsides. The fever is so lethal, it kills about a third of untreated victims, who are mostly farm workers, within a fortnight.

When other species drop out, an opportunistic mouse increases reproduction; its survival rate shoots up and so does the risk of transmission to humans. Overcrowding leads to fights. The fast-living deer mouse may seem shy to us, but males will chase and attack each other when they live too close together: the virus passes quickly through the population as rough-and-tumble fights increase.

It's not only the fighting—community nesting, sleeping together in cold climates, can speed transmission as well.[28] These patterns of horizontal transmission—through contact rather than through birth or nursing—have been found for lots of rodent pathogens, including Sin Nombre, the Seoul virus in Norway rats, the Black Creek Canal virus in cotton mice, and the Junín virus in corn mice. About half of the known rodent viruses affect humans.[29]

Several landmark papers have shown that high mammalian diversity can reduce the risk of contracting hantavirus. In one, Laurie Dizney and colleagues at Portland State University in Oregon set out live traps in city parks. Deer mice, the carriers of Sin Nombre, were the most common mammals at all the sites; the parks with the highest mammalian diversity had the lowest infection rates. Surrounded by predators and competitors, the deer mice stuck closer to home, lowering the number of their encounters and keeping the rate of transmission in check. When species richness approached zero, hantavirus increased exponentially.[30]

At the end of my trip to Millbrook, Oggenfuss and I crossed the parking lot and entered the emerald stare of the woods. We carried a corduroy cloth, weighted at one end, white as an unpainted canvas.

"When the ticks are questing," Kelly told me, "they hold on to the vegetation with six legs and reach up with their front two legs." Her fingers grabbed the air like claws.

We dragged the cloth along a 30-meter transect of the understory, mostly invasive Japanese barberry and garlic mustard, then she hung up the square meter on a broken tree limb. It made an abstract painting, white on white. We examined it closely.

"You're looking for specks of dirt with legs. Freckles with legs."

As we made our way across the canvas, we found three larvae. Oggenfuss told me that the best way to estimate tick density (and the burden to humans) is by walking though the forest—in white corduroy chaps: "man bait." Collecting whale feces may be rated one of the worst jobs in science, but at least you're out on the water, following whales with a plankton net, as if you were chasing butterflies. And if you're lucky, there's a Rottweiler, medicated for seasickness, by your side, urging you to follow his nose. But dragging your ass through the tall grass in white chaps trying to attract ticks—when it comes to bad jobs, man bait beats poop collector hands down.

"Do you really love ticks?" I asked Oggenfuss.

"Not at all." Assaying the ticks for *Borrelia* infection means crushing the exoskeleton of each and every nymph. "I feel good when I hear those things pop."

Oggenfuss turned back to her canvas. We dragged up ten ticks in a couple of transects. Focusing on this white flag, the ticks and their white-footed hosts, I started to lose sight of the trees. Yet as Ostfeld had made clear, Lyme disease is much more than a tale of ticks and mice and deer. "All of the animals matter, the entire community of terrestrial mammals and birds."

But what about Durland Fish's comment that if you wiped out everything, you'd have fewer diseases? "Yes," Ostfeld said, "if we paved over Dutchess County tomorrow, we wouldn't have Lyme disease. But if we paved over Dutchess County, no one would want to live here."

15

One Saturday morning, I rummaged through my daughter's room, trying to distract her with a tennis ball and a baseball cap way too big for her. There were stuffed triangles and cubes, a rattle, and a toy fashioned from a plastic water bottle and swizzle sticks from her native Vietnam. Ngan (pronounced like the element, neon) had a couple of imaginary beings, a blue Cookie Monster, an odd catlike teething ring from Thai Air. But most of her toys came with an evolutionary backstory. I sorted her toy chest phylogenetically.

If we were living at the end of the Age of Mammals, it was hard to tell that here. There was one Lagomorpha, a white rabbit. One Rodentia, a beanbag beaver. Two Primates, of course, human baby dolls with hats covering their baldness, a species largely ignored so far by my two-year-old. A few Carnivora: one pink dog, one white cat, one beanbag bear, a realistic skunk, and a cardboard leopard. There were several Artiodactyla: one white lamb blanket; one giraffe pull toy; one pink hippo; one purple hippo pull toy; and one buffalo, a gift from Yellowstone that she was fond of kissing on the snout. There were four Cetacea: one blue whale puppet with calf and two stuffed dolphins. And there were three Perrisodactyla: one zebra and two red rhino mittens, and one Proboscidea, an elephant pull toy. It's a big chest.

The rest of the Vertebrata were pretty sparse. Aves: one stuffed owl that hoots, two rubber ducks, one parrot puppet, three stuffed penguins. Reptilia: one boa. The Amphibia: one zip-along frog.

Once we left the Vertebrata, things got downright desolate. For the entire phylum Mollusca, with more than a quarter million living species, one octopus puppet. The phylum Echinodermata: one zip-along starfish and a plastic sea star bath toy. Of all the Arthropoda, which encompasses about four-fifths of animal species, one bubbling crab, with big blue neotenic eyes; and, perhaps, my favorite of all: one isopod *(Armadillidium ikeansis)* souvenir from a trip to Montreal. *Isopod* or "apoda" was one of my daughter's first hundred words. Unzip the roly-poly's mouth and you find a gray rat (score one more for the Rodentia). Both the isopod and rat are invasive here in the United States, but not without charm in their plush-toy form.

But there was something worrisome here. Except for a few marine creatures, my daughter's animals were almost all vertebrates, furry mammals, placental at that (no marsupials or monotremes). And, of course, there were no stuffed trees or plastic flowers. Do we not see them outdoors—or are they just not huggable enough? There were no archaea, no bacteria, not even fungi, a not-too-distant branch from us on the tree of life. One of my biology professors, Colleen Cavanaugh, had once told me that she longed to promote the life-giving joy of bacteria and other microorganisms in children's books or as stuffed "animals." Just think of a cuddly *E. coli*—gram-negative, lab workhorse, aerobic or anaerobic as the oxygen situation required. After Cavanaugh had extolled the virtues of a huggable microbe, one student quipped: "Wouldn't that just be a pillow?"

How did my daughter's toy chest compare to the endangered species list? There are currently 577 animal species and 795 plants listed in the United States. (There were also 587 foreign species, all but 3 of them animals.)[1] There was no need for a chi square test: although the toy chest and the list showed an inordinate fondness for chordates, Ngan's list was heavily weighted toward mammals (21, not including the baby dolls, with only 14 spread out across the other classes and phyla); on the endangered species list, mammals (83 species) were outnumbered by birds (92), fish (140), and even mollusks (104). The species are listed alphabetically—essentially arbitrarily from a biological perspective—within each taxonomic category.

You'd think such an evenhanded approach would help encourage equal protection under the law.

Unfortunately, funding isn't so equitable. In 2007, nine of the top ten funding winners were fish, receiving about $290 million. The only nonfish, the Steller's sea lion, was at the top of the list because of its impact on fish. According to one Alaskan biologist, the money had been a gift from "Uncle Ted." By most accounts, Senator Ted Stevens had little interest in the declining pinniped; he poured money into the research program to help biologists find a way to save his state's fishermen, who were worried that a government management plan would impact their ability to harvest pollack and other groundfish. The sea lion received more than $53 million in 2007. The majority of species got less than $50,000 each; many were lucky to get $20,000, barely enough to cover the reports that were due every five years. All of the 759 flowering plants on the list received less than $50 million combined.

If we remove the well-funded food fish, then things settle out about as we might expect. Mammals got $154 million (more than a third of that going to the sea lion), birds $130 million, reptiles and amphibians $59 million. All the inverts—the clams, insects, arachnids, crustaceans, corals, and snails—shared about $37 million. The ferns got about $700,000. Lichens accounted for little more than a rounding error, $112,000.[2]

The thing is, money can buy you recovery. Species that received more than a million dollars a year often saw improvements. Except for two beetles, a fairy shrimp, the Karner blue butterfly, and the flowering golden paintbrush, all 106 of them were vertebrates. This is one of the great weaknesses of the list and of the Act: we're stuck in a childlike view of the world, where furry animals—ones that resemble us or that we can fish and hunt—grab our attention, while the multilegged, sedentary, burrowing, nettled, tiny, or slimy are left out in the economic cold. Despite their enormous value, endangered trees and plants don't get the attention of the more charismatic animals (not even in this book). Ever heard of the false gumwood? It could soon be the dodo of the plant world. Found only on the remote island of Saint Helena, the diminutive tree, once collected for fuel wood and cleared for flax plantations, survives in just two tiny isolated populations: one of these has seven plants; the other consists of a single individual hanging onto a cliff. Self-incompatibility—a mechanism that many plants have developed to prevent self-pollination—has foiled attempts to reestablish it.[3]

Andy Dobson and colleagues at Princeton have suggested that the Fish and Wildlife Service's focus on animals may be misplaced. Protecting plants and other range-restricted species could be the most effective way of reducing species extinctions and the impact of human activities.[4] As plants go extinct, we lose biomass. Ecosystems with mixed species are almost twice as productive as monocultures. It's not just the highly productive species that matter; suites of complementary species are essential to facilitate growth and maintain specialized niches.[5] Despite the clear need for primary producers and diversity, we don't spend an awful lot of time or money protecting plants.

Paul Ferraro at Georgia State University and colleagues have contended that species showed no signs of recovery or declined further after they were listed if they were not backed by substantial government funds. Money works, perhaps because with funding comes enforcement.[6] The trouble is that, as Dale Goble told me, "Species recovery is funded at about twenty cents on the dollar, and an awful lot of money flows through earmarks to charismatic megafauna." If all species are of equal value, we need to be more equitable in the distribution of funds.

As I discovered in my daughter's toy box, kids are familiar with the lovable species, the furry and sociable, creatures with large eyes and expressive faces: chimps, dolphins, and dogs. Pokemon and Hello Kitty morph from there. We project our emotions onto these animals—lions are kingly, hyenas cowardly, even though they both hunt and feast on carrion. Traditional hunting societies like the

Cheyenne and other tribes of the Great Plains admired the courage of wolves, although ranchers saw them as thieves and cowards.

Up against genes, culture, and fashion, the best that education can do is extend the range of appreciation beyond our tiny twig to the furthest reaches of the phylogenetic tree. "To protect the nature that is all around us," environmental historian William Cronon wrote, "we must think long and hard about the nature we carry inside our heads."[7] Children can like stinging nettles and ants just because they're weird, though truth be told, as my daughter got older, she told me that she didn't like the isopod. "Too scary." She asked me to banish it from the toy chest. Out in the open, rather than stuffed into the bin with its antenna reaching out from under the lid, it presented less of a threat.

It may come down to the naming of things. Research in Switzerland showed that nature education fostered an appreciation of local animals and plants. The more wild species kids could name, the more they treasured them.[8] "Children are irrepressible taxonomizers, placing the world of distinct individuals into categories based on their appearance, their patterns of movement and their presumed deeper natures," Yale psychology professor Paul Bloom has written. "We are natural-born zoologists and botanists."[9]

Another one of my daughter's first words was *whale*. She got the humpbacks and the narwhals pretty quickly, but what most caught her attention was the tail. She pointed to a photo of a bald man in a hammock, hands outstretched, dark legs touching the sand: "Whale." It really did look like the fluke of a diving cetacean. A man in a black wetsuit stretched out on a surfboard, his shoulders humped like blowholes: "Whale." Like Melville, who saw a sperm whale in the ridgeline of Mount Greylock as he was writing *Moby Dick,* she started seeing whales everywhere. The streetlights diving into the PriceChopper parking lot? An upside-down teddy bear, his feet in the air? The fireplace tongs? Whales. Making these connections, developing metaphors, is probably a necessity when you only have a couple of dozen words to choose from. As her vocabulary expanded, she added *truck* and *tractor*—"Modern megafauna?" a friend asked—then *moo, neigh,* and *heehaw*. When she first saw a picture of Winston Churchill, she thought for a moment. Where had she seen this shape before? "Owl." She nodded in confirmation.

In developing a relationship with a particular animal and plant, you can learn to appreciate the whole ecosystem. The line between humans and animals has been firm in the Western tradition, even as the boundaries have shifted: man was the thinking, tool-making, selfless warrior—but it turns out that chimps, birds, and even octopuses and crabs use tools, wolves and lions kill each other in battles, whales exhibit altruism, and many species have consciousness. (Donald Griffin told me that when his influential *Animal Minds* was first published, it was treated like *The Satanic Verses* of animal behavior; zoologists have for the most part come around.) In the Aboriginal worldview, the boundary between people and animals was fluid, if it existed at all. Each person descended from animal ancestors, and each was indistinguishable from them; a man with kangaroo ancestors was a kangaroo. Lineages extended beyond animals: a person's totem could be a plant, rock, even water or smoke.[10] We don't need to go so far as convert to totemism, but Scott McVay, the executive director of the Geraldine R. Dodge Foundation, has wondered, "What would happen if every elementary schoolchild chose a creature, whether an ant, a bee, cricket, dragonfly, spider, waterstrider, snake, frog, fly, beetle, or bat, to study and report on repeatedly during his or her first six years of school? The capacity for bioaffiliation in the rising generation would be boundless."[11]

Ngan and I walked up the dirt road through the hollow. It had been a disappointing day—a rejection for a grant proposal, following bad reviews on a paper earlier in the week. Well, exercise reduces blood pressure, increases self-esteem, and lifts mood, right? Jules Pretty and colleagues at the University of Essex in England have shown treadmill exercisers some PowerPoint slides of rural and urban scenes as they worked out. (Just the thought of watching a Power-Point presentation during exercise makes me want to jump through a plate glass window.) Those who had seen pleasant scenes—pastureland or a view of the Charles River in Boston—showed significantly greater health benefits than those who had been shown unpleasant scenes or no slides at all. A ravaged rural environment—cars rusting in a field—had depressed their mood, reducing the benefits of exercise to dejection and fatigue. (Surprisingly, images of broken windows and urban blight had little impact, perhaps because we have come to expect cities to exhibit some decay.)[12]

Other studies have shown that images of natural scenes, technological nature, aren't enough; only real greenery and real views reduce stress. High-definition nature was no more restorative than a blank wall.[13] This is a problem for present and future generations, as we transition from actual experiences of the natural world to technological simulations. Several psychologists have proposed that generational environmental amnesia (biologists call it the shifting baseline syndrome) is a normal experience. Each generation takes the condition of its childhood as normal, no matter how much the environment has degraded over earlier decades and centuries.[14]

British researchers Richard Mitchell and Frank Popham found that circulatory diseases in people in poor urban areas were lower in the areas with the most green space. Parks invited exercise; they lowered blood pressure and enhanced healing. Stress-related deaths went down. The presence of natural areas reduced the inequalities in health between the rich and poor.[15] The psychological benefits of such green spaces increase with diversity. Researchers in a former steel-producing town in northern England have looked at species richness of butterflies, plants, and birds in different parks. Park-goers had stronger emotional ties to green spaces with higher bird diversity. Plant and habitat diversity made open spaces more conducive to reflection. As half the people in the world now live in urban areas, city parks, rich with native animals and plants, could enhance biodiversity, ecosystem services, and human well-being.[16]

Contact with nature may be one of the requirements of good health for children. Richard Louv, author of *Last Child in the Woods: Saving Our Children from Nature-Deficit Disorder,* has written con-

vincingly (if at times heavy-handedly) about the costs of losing nature, especially to children. The symptoms of what he terms "nature-
deficit disorder" are tunneled senses and feelings of isolation and
containment. Asphalt playgrounds encourage short, interrupted play,
not unlike TV and computers. Two-dimensional technology flattens
creative engagement; it takes out physical risk. In natural playgrounds,
children invent sagas that carry on from day to day—Milne's Hundred
Acre Wood—gaining meaning and complexity across the seasons.[17]

Given the overwhelming benefits of outdoor play, Oliver Pergams at the University of Illinois at Chicago and Patricia Zaradic
have uncovered a disturbing trend: fewer Americans visit natural areas
and national parks now than they did before the 1990s. People no
longer hunt, hike, or fish as they once did. (I think of a NASA scientist in Werner Herzog's stunning *Wild Blue Yonder*. His vision of the
future: "The ideal environment might be something like a shopping
mall in space." Humans would become tourists on their home planet.
"The earth will become a national park.") What's keeping people inside now? All kinds of fears: fear of animals, fear of getting lost, fear
of disease, fear of the unknown—and that's just for starters.[18]

Back on our road, Ngan was afraid of the red hen when we got to
our neighbor's coop. The chicken guarded her eggs as if someday one
would actually hatch. We took three and walked over to the corral.
After a few months, we had gained Taco and Chico's trust. The two don-
keys were our closest neighbors. The farmer's son was busy in the sugar-
house, boiling maple syrup for the year. It was late in the season; he was
cooking syrup that tasted more and more of tree. The hollow, which
to me could seem so small, must have felt infinite to a two-year-old.

How would she remember the Sundays walking though the
copse behind our house? The occasional hikes in the mountains? The
indigo bunting we saw this morning, or the rise of lightning bugs
after a thunderstorm? A flock of grackles shot from the copse like a
flight of arrows among the maples leafing out. The sumac leaves gave
light to the breeze.

Two blue jays alighted in the mountain ash. When the acorns begin to fall, the jays will lift the mast to new territories, carrying acorns

away from the parental trees, caching them in bare soil, thus restoring strands of oak to isolated forest patches. They help restore ecosystems when fields go fallow. It isn't just trees the jays help out. Black bear, turkey, deer, raccoon, and, yes, the white-footed mouse depend on oaks and their acorns. The jays have helped trees move north before, back when the glaciers retreated from much of North America, and they'll likely be called on to do it again as warming continues.

These birds exist in what Michael Soulé and colleagues call "ecologically effective densities."[19] Presumably, this was once the case for many rare and endangered species. The millions of bison on the grasslands of North America before Europeans arrived were reduced to a few hundred and have been restored to about fifteen thousand in the wild. But there are no longer any mass migrations—and fewer of the carnivores that once depended on them.

Hunting records from the East Coast, as far as they can be trusted, show a land once rich in predators. The colonists were quick to institute bounties to rid the area of wolves and mountain lions. They decided that the best way to still the nerves of settlers was to clear large swaths of big animals. In 1760, two hundred settlers circled an area of about seven hundred square miles north of Harrisburg, Pennsylvania, shooting every creature they saw as they marched toward the center. They reported killing 41 mountain lions, 109 wolves, 112 foxes, 114 bobcats, 17 black bears, a white bear, 2 elk, 198 deer, 111 buffalo, an otter, 3 fishers, 12 wolverines, 3 beavers, and hundreds of animals too small to record. In *In the Condor's Shadow,* David Wilcove wondered whether the numbers were too high for predators— they seemed to outnumber their prey—but the records were consistent with what would have originally been a healthy and widespread predator community throughout the Northeast.[20]

The trouble with the Endangered Species Act is that it aims merely to take away the danger of extinction. Being secure, a mere "ghost of past populations" as two sea turtle biologists called it, isn't the same as being relevant.[21] There's a big difference between sustaining a population at a low but safe size and restoring a species' historical role as a major player, whether it be wolves and elk, sea turtles and sea grass, whales and forage fish. To set our sights on alternative stable states, we often need to look deep into the past, to find a time when ecosystems were more productive and resilient.

The Marine Mammal Protection Act has it right, with one of its goals being to insure that each species plays a significant functional role in its ecosystem. Anyone who has ever studied whales has probably, after miles of empty waves and only mirages on the horizon, found his or her mind wandering. What would historical whale numbers have been in the planet's oceans? Logbooks can help us estimate how many were rendered over the years, but except for a few species, the final tallies are rarely much higher than today's numbers.

Genetic diversity, as we've discovered, is another way to estimate population size. The greater the number of animals in a population, the more chances for mutation. If you find a DNA sequence that is not subject to selection (which typically restricts variation), you can use this diversity to estimate historical numbers. In many cases, population bottlenecks limit diversity after species have been exploited. That's true for North Atlantic right whales, which were reduced to just a few dozen individuals in the early twentieth century: they don't have enough diversity left to give us much information about the past, though some scientists say that their numbers were never large. Species such as humpbacks and fins have enormous diversity, indicating that their populations were once much larger than they are now. Some historians, and whaling nations, were not pleased with this study (which I completed for my doctoral thesis), but it gave me a clue as to just how far we were from restoring the oceans.[22]

The historical numbers for many species are largely just of academic interest. Ernest Thompson Seton estimated that there were 75 million bison on the plains; more recent studies set the numbers closer to 25 million.[23] Either way, we'd be fortunate to restore just 1 percent of those to the wild. Even if we did return wolves and mountain lions to the East, they won't be found by the dozens in an area north of Harrisburg as they were in 1760. But for whales, history matters. Several whaling nations insist that populations are now close enough to prehunting numbers that whales should be hunted once more. But as DNA and the whale pump showed, whales and their prey existed in far greater numbers in the past. Restoring this ecosystem would benefit everybody.

Ngan and I stopped by a puddle near our driveway. Back in June, it had harbored about a hundred American toad tadpoles—swimming punctuation marks, single quotes—laid before the summer rains. When a stretch of sunny days had threatened it, we did our best, adding a few gallons of well water each afternoon. But all we had to do was forget that for a couple of days and the pond turned to dust. Ngan asked about the tadpoles for weeks.

We walked over to the dairy barn. Fifty milkers were lined up in two rows, udders enormous; the stench of manure permeated our clothes. The mollusk-like milk machines tubed up through the loft and down to a refrigerator, where the milk churned until the truck arrived. At close range, the cows Ngan thought she adored made her a bit nervous. She was more interested in the twenty or so barn cats racing up and down the hayloft.

I don't know what memories will remain with her—she was at the edge of the amnesia phase of childhood—but the transcendent experiences of youth resonate for biologists, as for artists, for writers, for everyone. These "radioactive jewels," as one psychologist put it, emit energy across the years.[24] Such jewels are rarely national parks but often simply a patch of weeds by a sleeping porch or a first trip to the mountains from the plains. The power of these memories may be one cause for the indelible nature of the shifting baseline syndrome.

This particular jewel was encrusted with an awful lot of cow dung. When we got home, I tried to sneak some time in with my computer, encouraging Ngan to put her toys back in the chest. I came across a study done in the Alps—there's always a study—showing that kids who grew up on farms had fewer allergies; asthma was rare. The two likely reasons: raw milk and fecal flora. Exposure to endotoxins on the surface of gram-negative bacteria, those cuddly pillows, may be critical in protecting against allergies. The hygiene hypothesis suggests that bacterial products are important in the development of immune responses in the respiratory and digestive tract. With little exposure to a diversity of microbes, the immune system becomes imbalanced, allergic, asthmatic. Researchers measured endotoxin levels on children's mattresses in farmhouses and in urban homes. They were highest on farms and in homes near cows, horses, pigs, and their feces and microbes. Natural exposure to farm animals lowered the risks of hay fever and other ailments.[25]

And it turns out these microbes can supposedly put you in a better mood. Depression and anxiety are higher in patients with chronic inflammatory disorders like allergies. *Mycobacterium vaccae,* first isolated in—where else?—cow dung, may work as an antidepressant. Lab mice exposed to the common soil bacteria produced more serotonin, a chemical known to lift moods.[26] Dirt-diggers may also be healthier; serotonin can boost the immune system, and staphylococcus produces a chemical that reduces inflammation around wounds.[27]

But sometimes even a tractor load of manure, 56 species of North American herbivores and their predators resurrected from the Pleistocene, a full moon high over the Green Mountains, and a spotted salamander in the basement won't stop the wailing, the tantrums of a 2.33-year-old. Ngan was in no mood for work. I shut down the laptop. She threw the isopod across the room, then the buffalo. Time for a nap, Sweetie.

16

Clearcut a forest, kill off a coral reef, and you destroy a medicine chest. That's one of the longest-standing arguments for preserving biodiversity: saving species and genetic lineages protects chemical diversity. Three-quarters of all drugs for infectious diseases were discovered in nature. The African liana *Anistrocladus* produces michelamine B, a compound used to fight HIV. Many antiviral agents are derived from two compounds, spongouridine and spongothymidine, isolated from marine sponges collected off the Florida Keys and the Bahamas in the 1950s. Sixty percent of the anticancer drugs on the market come from natural sources, among them a compound produced by the marine bryozoan *Bugula neretina.* The venom from cone snails led to the development of Prialt, a drug a thousand times more potent than morphine that can reduce pain in AIDS and cancer patients. A common parasitic plant in Africa, *Cassytha filiformis,* has been used as a folk medicine to treat cancer and trypanosomiasis, African sleeping sickness transmitted by the tsetse fly. A compound from the plant shows clear anticancer activity; a synthetic derivative is used to treat Parkinson's disease. And what would you expect from a widespread parasite with the common names of devil's gut and love vine? It's used to treat erectile dysfunction and sexual arousal disorder.

After I got through with my daughter's toy chest, I opened the medicine cabinet, lined up our drugs on my desk, and rang up the Natural Products Division at the National Cancer Institute. When I asked David Newman, the head of it, about our prescriptions, some from a recent trip to the tropics, he ran down the list at a sprinter's pace. But then, he has carved a career out of exploring the origins of the world's drugs.

Ciprofloxacin was a synthetic antibacterial. "There is a natural product in its background, but it's a rather amusing one," he told me. "The people were trying to make an antimalarial"—a protozoan— "and they happened to do an antibacterial assay. That was the start of Ciproflox. They were synthesizing one thing and came up with something quite different.

"Doxycycline," Newman continued, "is a very subtle modification of a natural product, tetracycline." The botanist Benjamin Dugger isolated the latter in 1948, when he found the compound in a bacteria common in the soil around cemeteries. It was the first broad-spectrum antibiotic, effective against gram-positive and gram-negative bacteria as well as fungi and protozoan parasites. His discovery sparked the development of numerous chemically altered antibiotics, but it wasn't new.

Tetracycline has been found in the bones of Nubian and Egyptian mummies. Where did the Nubians get it? Not by visiting tombs, but by drinking beer—the ancient brew was souplike, low in alcohol, and highly nutritious. Beer is a harsh environment for a bacterium; tetracycline was a byproduct of the microbe's stress. Did the Nubians appreciate the potential health benefits of the brew? Apparently so: beyond drinking it, they used it for mouthwash (tetracycline is still used for gingivitis), enemas, and vaginal douches, as well as a fumigant for anal disease.[1] And all I had been looking for was a prophylactic against malaria. Now my capsules, despite their Egyptian-blue glaze, seemed a little tame. The antibiotic is so potent, it is still effective against bacteria 1,600 years on.

Clavamox, prescribed for our cat after a skin infection, was a compound of clavulanate, a free acid that came directly from a microbe, and amoxicillin, derived from penicillin. Penicillin came from the mold *Penicillium notatum,* which infected a culture of staphylococcus in Alexander Fleming's lab in the 1920s. This "chance observation," as Fleming called it, led to the discovery of the natural pharmaceutical that has saved the most human lives. Mass-produced in the 1940s, it fought staphylococcus infections, syphilis, gonorrhea, tuberculosis, and other bacterial diseases. Penicillin, like many other natural compounds, is the scaffold for new drugs. Since its discovery, more than twenty thousand molecules have been synthesized from cephalosporin, a product that was first isolated from the sewers of Sardinia in 1948. It'll take more than a rain forest to save the world.

Tylenol, Benadryl, and Sudafed are all synthetics, Newman said. Aspirin, the most widely used drug of all time, was the only slam dunk—extracts of *Salix,* the willow, or *Spiraea ulmaria,* the meadowsweet, have been used to treat fever and pain since Hippocrates.

I was disappointed. My drugs all seemed to come down to bacteria and other invisible prokaryotes. Even the fungi were often nothing more than single-celled microbes. Did I just have the wrong drugs in my house? Sure, several of the drugs had natural origins, but they weren't exactly forest savers. Why protect biodiversity if two of the world's most important drugs were isolated in a sewer and a cemetery?

I asked Newman about the travel vaccines I had been prescribed. "All of these are basically derived from natural sources. In the case of typhoid, they raise a vaccine by adding the sugars that are on the microbe that causes typhoid, and they come up with a neurological response in a rabbit. Sometimes it can be in a horse, a rabbit, sometimes a chicken, sometimes an egg. It's an antibody response to the insult that you've given to the animal. They absorb them onto an adjuvant, which is just something that gives it a better response in your body, and they eject it. So you can say all vaccines have a natural source in their background."

Okay, so common barnyard animals are important. But how did I know that these vaccines were safe? Horseshoe crabs. Limulus ameobocyte lysate, discovered in the 1950s, is extracted from the blood cells of these marine arthropods. An essential part of their immune system, the enzyme binds and clots toxins produced by bacteria and other microbes. The limulus test—the lysate causes a clot when exposed to an endotoxin—can detect impurities in pharmaceutical products, plasma, serum, urine, blood, and medical devices. It's even used to detect bacteria in the International Space Station, helping to reduce the microbes we might bring into space or onto other planets and to track down stowaways before they hitch back to Earth. The ameobocytes are still harvested in the wild: each year about 250,000 horseshoe crabs are gathered on the shorelines of Cape Cod, Maryland, and South Carolina. They're bled in the lab—about a third of the blood is removed—and released back into the ocean. (About 10 or 15 percent may die as a result, although the lysate industry says it's much lower than that.) Here's a living fossil essential to human health.

My shelves were so depauperate, and Newman's knowledge so encyclopedic, I asked him about a few of the drugs that my grandmother, 98, had in her medicine chest. "Lipitor is a lovely one. They took the warhead of Nevacor, a hydroxyl acid, directly from a fungus, and they simply added different grease to the acid." The grease allowed the compound entry into membranes, where it decreased cholesterol. It was a natural warhead with a modified delivery system.

"Lisinipril is an ace inhibitor derived from a natural product. Amusingly, it's a snake peptide." Now we were getting somewhere. A snake, a horseshoe crab, a willow.

Newman breaks down the natural world into three categories: the microbial products that come straight from nature, the modified microbial ones, and the plant-sourced ones. (Animal products are too rare to count, I suppose.) When you talk to him, life and its myriad compounds eventually come down to the lesser flea (the one on the little flea on the back of a great flea and so on).

One of the most celebrated examples of medicinal products from plants is Taxol. The Pacific yew, *Taxus brevifolia,* had already been valuable for centuries—for bows and paddles, boat decking and furniture. The Okanagan people ground the wood with fish oil and used the dye to make red paint. Collected in 1962 by a USDA botanist working with the National Cancer Institute, the yew yielded a crimson liquid that showed cytotoxic activity. The two agencies went on to collaborate for more than twenty years, collecting 15,000 plant species and 115,000 extracts, in search of potential anticancer agents.

In the late 1970s, taxol was shown to arrest cell division in two slow-growing and difficult-to-treat tumors in mice: melanoma and breast cancer. A researcher at the Albert Einstein School of Medicine in New York observed that the extract inhibited the cells microtubules. Without functioning microtubules, chromosomes can't align and divide—cells can't reproduce. No growth, no tumors. The National Cancer Institute, faced with insufficient funds to extract and purify the compound, decided to deliver taxol and its data to a private pharmaceutical company in exchange for the development of a marketable product. When roads in the national forests where the Pacific yew grew had to be closed due to high fire risk, researchers couldn't get to the trees, the bark harvested was of poor quality, and the first trials had to be suspended. Even in the best of years, the Pacific yew was unlikely to meet the demand for the drug, as stripping the bark—the source of the crystals that formed taxol—killed the trees. The Environmental Defense Fund grew so concerned that in 1990 it petitioned for the yew to be listed as a threatened species, recognizing that the demand for the drug would be overwhelming. It was looking like a public relations nightmare: environmentalists taking a stand against cancer cures. Concern for the plant and for the drug supply waned after biochemists in France successfully synthesized the compound using taxane, found in the common yew. The multinational pharmaceutical company Bristol-Meyers Squibb registered the trademark Taxol in 1992. By 1998, it was the best-selling anticancer drug in the world, with sales of $1.2 billion per year.[2]

It's a lovely story, Newman admitted, but it's not quite true. "The people working on it were botanists. So when Gary Stroble," a plant pathologist, "found taxol-producing fungi in an enormous number of species, he was decried as a madman. Everyone knew it came from the plant."

You're telling me it isn't a plant product? I asked Newman.

"Exactly. This is the only time I've seen Norm Farnsworth," a research professor at the University of Illinois, who specializes in ethnopharmacology, "actually speechless. When I give a talk, Norm will usually say, 'Newman, that's the biggest load of shit I've seen in a long time.' This is a standard greeting from Norm. I was at a meeting a couple of years ago where I said, 'Now, let's think of it in a slightly different way. If you're a plant, and you're being attacked by fungi—those are the main attackers, not chainsaws—and you've got to postulate the following: the signal tells the central genome of the plant, wherever that is, *I'm under attack.* The prevailing thought is that the plant produces a toxin, such as taxol, to kill the fungus. Then the plant has to get rid of the toxin, since it could also kill the plant. So that's why you find the toxin in the bark—it's the one way out. But let's postulate something a little different. You've got fungi all over the tree that have the potential to switch taxol production on and off. . . . The signal goes to the central genome—and I'm not defining what that is because no one's ever done so—and the signal goes back to the area where it's under attack and says, *Switch on taxol,* to the fungus. The fungus switches on taxol, effectively killing itself and the twig or leaf that it's in, on, or under, but nowhere else do you have to worry. Absolute and utter silence from a group of very competent botanists and chemists."

Some in the pharmaceutical industry had known since the 1960s that taxol and other plant products were really secondary metabolites from fungi. "They were trade secrets within the industry," Newman said. "I mean, it was a black art—how do you persuade a microbe to switch on or switch off? It still is."

Speaking of black arts, I asked Newman about the origins of Viagra. "Viagra is an amusing story. Originally it was designed for cardiovascular disease, based upon a natural product related to caffeine. It failed as a CV drug, but no male on the clinical trials wanted to give it up, because of its effect on erections. Caused significant priapism in most. The CV pharmacologist at Pfizer UK then had to

spend almost a year convincing his bosses that (A) there was a disease that could be recognized, now known as "erectile dysfunction,"
and (B) that they had a treatment for it. It became the fastest drug
to do a billion dollars in sales, ever."

Viagra soon challenged traditional Chinese medicine, where
the treatment for erectile dysfunction had been the consumption of
some or all of the parts of sea cucumbers, pipefishes, sea horses,
geckos, deer, seals, rhinoceroses, and tigers. Frank von Hippel, a
biologist in Alaska, and his colleagues interviewed men over 50 in
Hong Kong; overwhelmingly they had switched from Chinese to
Western medicine to treat this problem, even as they held on to traditional treatments for other ailments. There isn't good data on the
black market, but the legal sale of deer antlers and seal penises
plunged following the 1998 release of Viagra. It's Ecological Economics 101: eliminating market forces can often be the best conservation solution.[3]

I had read that 80 percent of all natural pharmaceutical
products—a $65-billion-dollar-a-year industry—was from plants.
"Ah, now this is where we get the ethnobotanists upset," Newman
responded, "because they keep saying, 'Oh, but look, 80 percent of
the world uses plants as their source of medicaments.' Well, that's
perfectly okay. But if you take the Chinese *Materia Medica* and read
it carefully, what you find is something very interesting." The *Materia Medica* is one of the oldest surviving compendiums of drugs, including more than 52 prescriptions dating back to 1100 B.C. Newman was a walking *Materia Medica*. Although I had sent him a list
of my medicines in advance, he had, as far as I could tell, just reeled
off their sources as we'd discussed them on the phone. "There's a
strict definition of where you get the plants, what conditions they are
grown in, the climatic times to pick, and the altitudes."

It's not just the plant but the conditions for the microbial flora to
produce the right compounds. A medicinal plant in one country
might not be curative in another. Some microbes depend on multi-
cellular plants and animals to provide their ecosystems—a root sys-

tem or a cow's gut—and we need the compounds that the microbes provide. It wasn't enough to save the species. You had to save the habitat, and even the microbes that can change with temperature, humidity, and soil: the *terroir*. As with endangered species protection, diversity and their habitats mattered.

"It gets to be a little tricky," Newman admitted, "because the argument the botanists have used against it is that, oh, when you isolate the fungi, you only get a tiny bit of material. The rest is just carried over during the isolation process. Well, that's absolute bullshit, and the reason is that 99.9 percent of botanists have no knowledge of microbiology or of microbial genetics. . . . A lot of them have spent their lives defining taxonomy as a result of the molecules that they found in the plants. Well, if it turns out that that molecule is not produced by that plant, or is only produced by that plant when particular microbes are present, oh, do you end up with people being very upset! You just destroyed their scientific work."

Invertebrate biologists weren't safe, either. Aplidine, shown to shrink tumors in pancreatic cancer, has been isolated from tunicates, filter feeders known as sea squirts that feed on algae. "What they found was that the alga contained the compound," Newman bragged. "The tunicate was simply absorbing it. But it gets better. They've now isolated a vibrio," a marine bacterium, "from the alga that produces the compound during fermentation." Newman ran through several more examples with glee: the antiretroviral camptothecin, from the Asian tree *Camptotheca acuminata,* shown to fight HIV; podophyllotoxin from the may apple, used against genital warts; even the alkaloids from the Madagascan rosy periwinkle (endangered in the wild) used to fight leukemia and Hodgkin's disease. Every one of those, according to Newman, has been shown to be produced by fungi that live on or inside the plant. Most fungal species can produce dozens of compounds, but they only make a particular chemical when they need it. What researchers need, said Newman, is a promoter. "We're looking for a Don King to put in front of these genomes—as long as we can control it. We don't want a Mike Tyson coming out the other end. We want a Golden Gloves, but somebody whom we can control."

When I had visited Yellowstone, I started feeling sheepish about spending so much time with all that megafauna—coyotes, buffalo, bears, wolves—and decided to visit one of the park's most valuable species. I drove down to a small spring off the Firehole River. Unlike the Lamar Valley, there were no spotting scopes, no daily observers, no crossing guards to help me locate the creatures. When I asked at the information station, I couldn't find anyone who had a clue how to find Mushroom Spring, the type locality for the species I wanted to see. Someone eventually dusted off an out-of-print book that gave a rough location: one hundred yards east of White Dome Geyser.

It was getting late when I arrived, the sun drifting toward the horizon. I followed my shadow east, skirting several pools; steam rose from the blue and green caverns, each with a hardened edge like a coral atoll or a salt-rimmed margarita glass. Some trees had fallen over the landscape. Some stood calcified, recently deceased, their tops knobbed like narwhal horns. Several wore a few brown needles like a tattered woolen dress. In the pools, the fallen branches looked like bleached antlers surrounded by fish bones of old needles. Then I noticed where the real action was: at the bottom of the pool, rusty orange algal mats, as thick as woven blankets torn in places and rumpled like an unkempt bed.

It was the spring I had come looking for. Minute bubbles rose to the surface from the chasms—one dark green, one mustard yellow—brightening the water for a moment, then spreading like scattered raindrops before a storm. Every once in a while, the chasms bubbled over, like a leviathan rising to the surface to take a quick exhale, then disappeared. Off in the distance, White Dome Geyser spouted a slender column 60 feet in the air.

The metazoans—bison, ravens, elk, and insects—even the trees and most fungi merely skirted the edges of this pool. But on the algal—on the microbial—level, the spring is, well, a hotbed of activity. As Newman made clear, bacteria, protozoa, and viruses are at the heart of biodiversity, even if we don't see them. Everything on the endangered species list—vertebrates, invertebrates, flowering and nonflowering plants (such as ferns, conifers, and cycads)—are found on one slender branch of the tree of life.

The two lichens on the list, the Florida perforate cladonis and the rock gnome lichen, extend the taxonomic variability by the barest of margins to the kingdom of the fungi. (Lichens are chimeras: part fungus, part alga. I imagine Newman would stress that several of them—the nitrogen fixers—are part cyanobacteria.) If you looked down on the crown of the tree of life from above, the listed species and all their relatives, including us, would spread from one o'clock all the way to about two-thirty.

In 1965, Thomas Brock, a microbiologist at Indiana University who had begun studying photosynthetic microorganisms along the outflow channels of these springs, noticed that there were masses of stringy, or filamentous, bacteria in waters that were much hotter than those where his photosynthetic organisms survived. Before leaving their research station, he and his students collected the microbes, along with some algal mats from Octopus Spring (temperature approximately 185°F). As he crouched at the edge of the mustard yellow pool, perhaps warmed by the mist as the breezes blew in from the west, he could not have known that this collection would help change microbiology, medicine, and even the study and protection of endangered species.

Initial attempts to culture the pink bacteria failed. The following year, Brock moved to the more temperate Mushroom Spring (156°F), scraping off a bit of the underlying sinter. In October 1966, his lab cultured a brand new archaean thermophile, which he later named *Thermus aquaticus*. At the time, this discovery was of interest to just a few bacteriologists. But then Kary Mullis at Cetus Corporation and colleagues figured out how to use an enzyme isolated from the microbe, Taq polymerase, to copy single strands of DNA. All organisms contain polymerase, a molecule that forms and repairs DNA. Taq polymerase worked at temperatures as high as 165°F, the point at which DNA is denatured into single strands. By adding nucleotides and primers—small pieces of DNA that kick-start synthesis—to the mix, polymerase could make copies of a desired sequence. Once you isolated DNA, you could copy a desired sequence over and over, exponentially.

The rest is history. *Science* magazine declared Taq polymerase the "molecule of the year" in 1989. (Three years earlier the journal had rejected Mullis's description of the polymerase chain reaction.) In 1991, the Swiss pharmaceutical company Hoffman LaRoche paid $300 million for the world rights. Yellowstone got nothing for the discovery; there's not even a sign indicating where *T. aquaticus* was found. Since then, microbes in national parks have been protected: researchers have to sign contracts guaranteeing that any commercial discoveries made in any sites where scientific research occurs will be shared with the Park Service.

Such single-celled microorganisms, known as Archaea, look and act like bacteria but have a deep and distinct evolutionary history. *Pyrodictium* forms microscopic ghost nets around the black smokers of deep-sea hydrothermal vents; far from light or oxygen, it thrives in temperatures well above the boiling point. *Halobacterium* flourishes in waters too salty for most organisms. Others live in petroleum deposits. Microbes are everywhere: there are symbionts, thermophiles, infectious agents—even endoliths that live in the tiny pores of rocks. "If you really want to blow people's minds," Newman had said, "tell them that there are one hundred thousand microbes in a milliliter of clear water. A million microbes, and you get a slight haze. But one hundred thousand per cubic centimeter is perfectly clear."

You might even find *Thermus aquaticus* in there. At the time of discovery, it was thought to be an exotic microbe in a remote cauldron on a large and active caldera. Brock later found the archaean in hot springs around the world. Then it showed up in the hot water system back in his lab at Indiana University. It's even in tap water. Related archaeans have been found in the human colon, vagina, and mouth. There are at least seven hundred species of bacteria living in your mouth (many detected with Taq).

How about that gut? Ninety percent of all the cells in the human body—slightly less that 2 percent of our body weight or about a kilogram—are bacterial. Much of this is *Lactobacillus,* lactic-acid bacteria, residing in the large intestine and other parts of the alimentary canal. People who dwell in rural parts of the developing world—Africa, India, Mexico—have more than twice as much health-promoting bacteria than the rest of us because they live on vegetables and fruit fibers, some of which are stored in the soil, where they ferment. Pozol (from corn, in Mexico), kimchi (from cabbage, in Korea), gundruk (from greens, in Nepal), peuyeum (from cassava, in Indonesia) are a few of the fermented foods that cultivate commensal flora. Our Paleolithic ancestors had a diet that contained at least a billion times more *Lactobacillus* than we consume.

I had been disappointed to find so few plants and animals in my medicine chest. I was certain my fridge, void of meat except for a little brown bat in the freezer (for science, not for consumption) would have at least a few higher life-forms tucked among the fungi and bacteria. (Sorry, Dr. Newman, if my phylogenetic bias is showing.) I'm not going to line them up, much as I like a good list. It's far easier to order takeout.

Humans regularly consume about seven thousand plant species, around eight hundred of which are critical to our diet. Almost all the essentials are angiosperms, which supply more than half our energy needs; in some cultures they provide as much as 90 percent. The rest of our caloric intake is comprised mostly of vertebrates, with the occasional fungus or alga thrown in.

We're not particularly picky eaters, since we can count on cooking and fermenting to knock out many toxins, parasites, and diseases.

Serban Proches and his South African colleagues found that our eating habits are rather random; we pick out fruits, seeds, stems, and roots from across the phylogenetic tree of about three hundred thousand flowering plants.[4] We are big fans of the Rosaceae, the rose family, and eat many of them raw: apples, plums, peaches, apricots, pears, strawberries, and brambles (raspberries and blackberries). These fleshy fruits evolved for seed dispersal by vertebrates, so it's no surprise that their edible charms—flashy colors, mouthwatering flavor in the raw—worked on us so well. Grasses, the family Poaceae, which includes wheat and rice; and then Fabaceae, beans and peas, play essential roles in human nutrition. They're almost always cooked—cooking reduces plant toxins, breaks down fiber, and lets us eat faster. Raw wild meat, rich in connective tissue and muscle fiber, slows intake down to a gnaw.[5]

Now, about that takeout. Want a Big Mac? With fries and coffee, it's a cornucopia of diversity, if not of nutrition or sustainability. To the familiar indices used to measure biodiversity—Simpson's, say, or the Shannon—we could add the Big Mac Index (devised by *The Economist* back in 1986, the index uses the price of a Big Mac around the world to compare international currencies to the dollar), given a twist by Proches and colleagues.

There's barley and wheat in the bun, from the family Poacea, grasses of Turkish (or Iranian) origin. Onions, in the lily family, also come from Turkey. Corn—in the buns, sauce, and oil—is a grass from Mexico. Chili pepper, from Mexico, too, is in the Solanaceae family, as are nightshades like tomatoes (Mexico) and Andean potatoes (fries are the best reason to keep McDonald's alive). There's tumeric (Zingiberaceae) from Southeast Asia and mustard (Brassicaceae) from India. In the sauce you'll find oil pressed from cotton seeds (Malvaceae, of multiple origins), and sugar, typically from beets, family Amaranthaceae, originally from the Mediterranean. Also from that region comes lettuce, family Asteraceae. Soybeans are everywhere, in the buns, sauce, seasoning, cheese, and fries; they're legumes, of course, in the family Fabaceae, their country of origin China. Coffee is a rubiacid, probably from the highlands of Ethiopia.

The entire diet of someone in a developing country typically contains about 50 to 100 species. You can order up twenty species, from twelve different families, in a single McMeal, no need to Super Size. A trip to the drive-through won't save the planet, but that paper bag holds several keys to human evolutionary history: our geographic range, our ability to digest the seemingly inedible—from coffee to chili peppers—and our hunger for the tree of life.

17

The second driest year on record had just ended at Lake Lanier in Georgia. Bright red mud and orange sand surrounded what water was left, like lips puckered to suck up the rest. Here and there, sailboats listed high and dry. The public boat ramps dropped off like concrete cliffs above murky green water, the three-dollar entrance fee suspended, a tollbooth covered by a black plastic bag. Abandoned in the mud, floating docks struck odd cubist angles, as if the reservoir were just a mirage, though some home-owners had built temporary docks whose new sections snaked toward the receding waterline. Islands lay anchored to the mainland, shore-line trees left hanging in the air with roots exposed. Muddy water—all that was left of a stream—dribbled under a bridge.

I drove over the Buford Dam, a large earth embankment with saddle dikes that wind for about a mile in each direction. When the hydroelectric dam was finished in 1956, it took three years to fill "Georgia's freshwater ocean." The lake was named for the nineteenth-century poet Sidney Lanier, who had written "The Song of the Chat-tahoochee" celebrating the river destroyed by the project. Lanier must be spinning like a turbine in his grave.

Downstream, the landscape looked almost natural: water rip-pled around red rocks, hardwoods leaned into the river, and fly fish-ermen worked the stream while Canada geese honked along the shore. But then I looked up. Like an Egyptian monument, the dam rose nineteen stories above what was left of the river. From a man-made canyon blasted through the rock, water emerged, funneled by a chain-link fence topped with razor wire. Power lines spread out from the dam, supplying electricity to metropolitan Atlanta. The river rocks were a little discolored above the waterline, exposing the reduction in flow in the last few years.

Under an agreement signed the previous year, the US Army Corps of Engineers was required to send enough water into the Chattahoochee and Apalachicola rivers to protect two endangered mussels—the fat threeridge, *Amblema neislerii,* and the purple bankclimber, *Elliptoideus sloatianus*—even though they lived two hundred miles downstream in the Florida Panhandle. The delivery of fresh water was also critical to the spawning grounds of the threatened Gulf sturgeon. Georgia politicians pleaded with the Corps to reduce the amount of water released into the Chattahoochee: in the midst of the drought, Atlanta, with more than five million residents but no natural water supply of its own, needed a drink.

Back before the lake was filled, Atlanta had been so small that it decided not to contribute to construction costs of the dam, a determination that would later weaken its claims to the water. Fifty years on, the city led the nation in population growth, adding more than seventy-seven thousand people each year, 55 acres of concrete, rooftops, and parking lots per day. In 1989, Georgia entered into an agreement with the Corps to double the amount of water taken from Lake Lanier. "Every drop counts," Steve Herrington, an aquatic biologist with The Nature Conservancy, told me. "It is a singular phenomenon to have a major city, the capital of the New South, with no water source." When you drive across the Chattahoochee in Atlanta these days, you hardly notice it: the river is so thin.

"Follow the water," Herrington said. "There was a lot of political squeezing going on, with the governor at the Corps office every other week. He even called my wife a meaningless bureaucrat, which I loved." (His wife worked for the Fish and Wildlife Service.) After the Corps acceded to Georgia and lowered the level, Florida and Alabama sued; the compact to share water that the states had agreed on was only as good as the rain. Things got so bad that Georgia lawmakers introduced legislation to move the state border north—to seize land that had been in Tennessee since 1818—all to capture a river. The governor turned to Plan B: a miracle. "O, Father, we acknowledge our wastefulness," Sonny Perdue prayed on the steps of the Georgia State Capitol in front of a crowd of about a hundred people. "We've come together here simply for one reason and one reason only: To very reverently and respectfully pray up a storm."

Mussels don't have a lot of friends, at least outside of the conservation community. By the end of 2007, when the Fish and Wildlife Service agreed to reduce the minimum flow by almost 20 percent, many mussels were already stranded, and much of the gravel habitat of the Gulf sturgeon was useless; a sudden further drop was likely to have a grave impact. The Service amended its proposals, suggesting a gradual reduction in water flow. But in February 2008, the US Court of Appeals ruled that it would take an act of Congress for Georgia to divert any more.[1]

For conservationists like Herrington, Perdue nailed the problem: much of the city's water intake was lost or wasted. "People are still watering their lawns. A million gallons a day are siphoned off to create snow in August for an indoor snow park." In July 2009, US District Court Judge Paul Magnuson found that the agreement between Georgia and the Corps to draw off more water was illegal. The dam was intended for hydroelectricity, he noted, not water supply.[2]

If you listened to Georgia politicians, this was another classic conflict pitting animal against human needs: Atlanta's citizens should override the Corps and the clams. Governor Perdue called it a choice "between mussels and drinking water for children," urging President Bush to declare the drought a major disaster, which would exempt the state from compliance with the Endangered Species Act. Congressman John Linder claimed that the Corps was behaving "as though mussels are more important than our children and grandchildren." The *Wall Street Journal* described the disagreement as a "War on a Little Bivalve."[3]

Would the purple bankclimber and the fat threeridge be the next snail darters, the new spotted owls? The Chattahoochee's waters would have to pass through fifteen dams to reach the gulf, a journey

of about eight days. It would take me about eight hours to drive to
Apalachicola from Lake Lanier. More reliable than the water flowing
out of Buford Dam was the traffic on I-85 heading south to Atlanta,
and then on the state roads through the Panhandle to the Gulf.

The days are for working, the nights for driving. As the night
drew on, the music on my iPod seeped in.

The Nights for Driving Music

You Were Always There
Lyle Lovett

In the Pines
The Louvin Brothers

Big River
Johnny Cash

Step Inside This House
Performed by Lyle Lovett

What Keeps Mankind Alive
Tom Waits

Many Rivers to Cross
Jimmy Cliff

Greenville
Lucinda Williams

Misery Is the River of the World
Tom Waits

Im Abendrot
Richard Strauss, performed by Janowitz and Karajan

One Hundred Years from Now
Graham Parsons, sung by Wilco

Pilgrim
Steve Earle and the Del McCoury Band

At play each day, as I drove through the South
(with apologies to Richard Long)

I do love a list—or better a tree of related species, a compilation of songs. What biologist doesn't? Can't get enough of ill-fated love affairs? Try preserving an at-risk species. I remember, when I worked as an editor in New York years ago, hearing one of my favorite radio DJs, who had just got her hands on the latest Endangered Species List, announce how shocked and sad its 44 single-spaced pages left her. I was inspired to write to the Department of the Interior for a copy of my own. Thirty pages of animals, fourteen of plants. I wanted to save them all.

Later, I would learn that the list was dwarfed by two others that were not as politically driven: the IUCN Red Data Books and the list compiled by NatureServe. Now E. O. Wilson and colleagues are racing to put an entire *Encyclopedia of Life* online, hoping to deliver, in less than a decade, basic information on every species, all 1.8 million of them, common and rare.

There are 297 freshwater mussel species in North America, separated into two families: the Unionidae and Margaritiferidae. Many of these bivalves reside in the Southeast, in rivers that drain into the Gulf of Mexico, in waterways as richly named as their bivalves: Suwannee, Apalachicola, Ochlockonee, Pascagoula, Pearl, Mississippi, Coosa, Neches, Little Brazos.

Jim Williams began working at the Endangered Species Office in DC back around day one. During the nearly three decades of his career—he later moved to the US Geological Service—almost all of his work was on rivers, much of it on Southeastern mussels and fish, including the snail darter. In a 1993 paper, Williams showed that more than two-thirds of all mussel species were threatened by invasive mollusks and by dams, runoff, channelization, and sedimentation. At least twenty thousand dams with a depth of at least twenty feet have been erected on US rivers and streams.[4] Perhaps hardest hit were the mollusks: only 70 mussel species are considered stable.[5] Thirty-seven mussel species have probably gone extinct in the United

States. On the Coosa River alone, 59 species of mussels and snails were annihilated following the construction of a series of dams in the early to mid-twentieth century.[6]

Williams's conclusion: the disappearance of mussels from American waterways is one of the most alarming extinction events in modern history. In the Mobile River drainage of Alabama, the southern pigtoe, Coosa moccasinshell, orange-nacre mucket, fine-lined pocket book, and seven other mussels were added to the list in 1993. They have wonderful names, but several are barely an inch long and hard to distinguish without a degree in invertebrate zoology. Even Williams, when I asked him, had trouble telling me how he knew one clam from another.

Early on, Williams and his colleagues knew there was going to be trouble listing mussels. They met at a bar to discuss priorities.

"We had to establish some rules. Is it bigger than a breadbox? No. Would you feel bad if you ran over it? Not really. Is it red, white, and blue? No. Sigh. What do you do with an animal that only has one foot and no head?"

They weren't manatees or bald eagles: they were living rocks. There was confusion over mussel systematics. Genetics, morphology, and the movement of bivalves between drainages, carried in ballast water or through canals, keep mussel taxonomy and conservation in flux. There are several hundred names for the freshwater mussels of France, but only seven recognized species. Estimates for the number of species worldwide range from 40,000 to more than 150,000.

"The Bush Administration has been worse than the early Reagan years," Williams told me. "They just said, 'The hell with it. We just won't list anything.' So now there's this huge backlog. The Service's biologists say, 'They don't give us any money to do listings.' When they do that, it just irritates the hell out of me. 'Cause if I'm sitting there drawing a damn paycheck, I've got a little spare time. What if I want to come in on weekends to do it just out of dedication and commitment? I can't do it because we aren't getting paid to do that? You gotta be kidding me! I volunteered a couple of years ago. I said, 'Look, you've got six or seven mussels over here in west Florida that are in pretty bad shape. Two of them are literally dangling by a thread. I will write the packages for you for free.' 'No, we can't do that. We can't do that.' This is just because the administration has said, 'We don't want any listings.'"

Williams and I drove past turf farms and cotton fields and into the Panhandle. He was taking me to see the Apalachicola's mussels. "We have to stop for tupelo honey," he said, as we approach Blountstown. "It's the best honey in the world. I got a lot of people hooked on it."

Williams is a tall, solidly built man with a gray mustache and gold-rimmed glasses. He was wearing shorts and a khaki shirt—James Beard in the subtropics, I thought for a moment. Sixty-seven and retired, he had been married and divorced twice. "Women sometimes interpret my interest and love for my work as putting them second," he said. "For the past ten years, I've had nothing to worry about but scheduling my own time."

A few years ago, as Williams's daughter approached graduation from high school, she sat him down and said, "Dad, we have to talk." A vortex of parental worries spun through his mind. Pregnancy? Okay, we can deal with that. Trouble in school? We can deal with that, too. "Dad," she confessed, "I'm going to study biology." There was no hope he said, and laughed. She's now a coral reef ecologist in the Florida Keys.

At a roadside stand, Williams bought ten pounds of "liquid gold." As it turned out, the honey was as tied to the fate of the Apalachicola as the mussels: the tupelo trees that provide the nectar grow only in the floodplains. When those flatlands get too dry, the trees disappear and so does the honey.

We drove across Dead Lakes, a stunning landscape of black water and cypress stumps that seemed part Amazon, part Asian woodblock. "There's a gathering storm out there that is just scaring the hell outta me right now," Williams said, "and that is this drought, especially in the Piedmont, because it has no groundwater. It's all either schist or granite or gneiss or some hard rock that has no water-storage capacity. It's like sitting on a concrete hat. In the upper part of the coastal plain, they're looking at damming up streams. Well, they know they can't dam up the Flint River, because people would just raise a whole lot of hell about that. So what they're gonna do is go in and dam up the Flint tributaries. What would happen if you started cutting off the blood vessels from your body or cutting off the arms and limbs of an animal? Well, let's chop off an arm here and a leg there. It's still alive but it's not going to get very far.

"We all know intuitively," Williams said, "any time you protect a river that flows into a bay where you've got all this marine production that that's an obvious benefit. But you get radio stations in Atlanta with 'Mussels Suck' T-shirts, or 'Kill the Sturgeons,' things like that. And that's only going to increase in the Southeast because growth has been so rapid in the past couple of decades. Even though we have a lot of resources, water's gotten rare."

We met our boatman, Jerry Ziewitz of the Fish and Wildlife Service, at Gaskin Park, a launch surrounded by fishing camps. A gray-bearded man in a straw hat, fishing with a long bamboo pole, told Williams he had caught a two-pound sheepshead upriver a couple of weeks ago. "All the saltwater stuff's moving up."

We boarded the Fish and Wildlife Service boat, equipped with a Mercury 150 outboard motor and a couple of clam rakes. The shoreline, littered with slab along the camps, gave way to yellow sandbars, a twenty-mile stretch that was the only place on Earth where the fat threeridge resided. Growing just bigger than a walnut, *Amblema neislerii* lived along the upper lip of the river channel and in sloughs and tributaries off the main channel. There were approximately 235,000 here before the waters receded in 2006, almost all found south of the Woodruff Dam, on the Florida-Georgia border. When water levels were dropped, thousands of mussels lay exposed in Swift

Slough; in hot, shallow water, all they could do was expel their larvae in a desperate attempt to breed before they died. About fifty thousand fat threeridges perished.[7]

Once every few thousand years, when one stream captures another, mussels can expand their range to a new drainage but, for the most part, mussels are their rivers. Consider the purple bankclimber, one of the two species that had Atlanta up in arms. The Apalachicola River was in the bankclimber's DNA; this mussel, at least when it was common, was central to the ecology of the river. Spawning fish thrive on the mussel's free-living larvae—many of these are consumed by the fish, but some settle on the gills or fins. Bankclimbers and threeridges, like most freshwater mussels, have a parasitic life stage; the larvae, known as glochidia, feed and develop on the swimming fish—the fish keeping the young mussels from being washed out to sea while they're developing—until they settle on the river bottom. The fish community (largemouth bass, sunfish, brim, and striped bass) was dependent on the flood plain; the mussels were dependent on the fish.

As we motored downstream, houseboats from primitive to prefab, in varying states of decay, dotted the banks. There was an occasional hog pen for hunting dogs. "They'll just sit here," Williams said, "until the next big storm comes. Then they'll end up in the trees."

We stopped at a slab of sand cutting through the forest. We had arrived at Swift Slough, a waterway in name only. There was no sign of a stream. I asked if the changes in the water level were natural. Ziewitz laughed. "Nothing's natural here."

About 20 percent of the entire population of fat threeridges once resided in this waterway, but after the Corps lowered the water flowing from the Chattahoochee, there was a massive die-off. Aside from the main stem of the Apalachicola, there wasn't much "natural" left. The water flowing over the dams had created deep channels, and the river became entrenched between the new banks, no longer reaching the floodplains. As water levels dropped, tree density in the bottomland forest declined by more than a third. The density of water tupelo, the dominant tree in the swamps, was down by about a fifth. Almost half of the Ogeechee tupelo trees, the primary source of tupelo honey, have disappeared since 1976.[8] (Even back then, the populations were no match for what Bartram described in the eighteenth century.)

"If you protect the freshwater mussels and Gulf sturgeon in the Apalachicola River," Williams said, "that in turn protects Apalachicola Bay and also the floodplain. Without the periodic regular flooding, we wouldn't get a lot of the trees blooming every year. When you disturb the hydrologic cycle, which they already have, then that's gonna affect flowering, which in turn affects production of honey. Now, most of the world could give a rat's ass about the flow of tupelo honey coming from Wewahitchka, but it means a lot to me." Without the honey, Williams wouldn't be stopping in Blountstown anymore, and the beekeepers would have to move on, or find another livelihood.

Williams mentioned that he was busy describing a new mussel species—the cypress floater, *Andonta hartfieldorum*—found in drainages from eastern Louisiana to western Florida. Why hadn't it been discovered before? "It lives in the worst habitat imaginable," he said, "backwater sloughs. You can't walk in the silt of those sloughs. You have to crawl in, dodging snakes and snappers and alligators, feeling with your hands for the mussel, since you can't see a thing in the murk." This was as close to rapture as Williams got, as far as I could tell, except perhaps when he got his hands on some tupelo honey.

"When you love the work," Williams said, "how can you quit when you retire?"

As we continued downriver, an enormous dune heaved up in front of us: dredge spoil from keeping the channel clear, the Corps having to re-dredge every few years as the river eroded the sand. The last time they had dredged was 2006; it proved too costly to ship the sand offshore. The spoil heap had been out of compliance for years, Williams told me. Another biologist said that the local officials understood the issues "but they're getting paychecked on from higher-ups." Nothing's natural here.

The River Styx, lined with Ogeechee tupelo and cypress trees, was where we stopped for lunch. Bright invasive vines reached out from the branches. A mullet jumped near the bank. "It's where you meet the boatman," Williams said.

He asked Ziewitz about the new listing packages for a couple of mussels on the nearby Florida drainages; some of them were hanging on by a byssal thread.

"Because of the lawsuits," Ziewitz answered, "we don't do listings." Williams and I exchanged a glance. "We don't really have the funding," he added.

"As you can see, it's funding," Williams said, somewhat drily. "There's a pretty small community of musselheads. So the service will get some divers in the water to collect them and culture them." Short of extinction, this was the musselhead's worst-case scenario: a population extinct in the wild and dependent on captive propagation. A living rock among the living dead.

We motored up to four young women in wetsuits and a male graduate student chin deep in the water, all of them proud of their new suction dredge, which had been developed for small-scale gold mining. Standing in the river, with the Apalachicola wicking up my linen shirt, I was close to the heart of the global hotspot of freshwater mussel diversity: there were 30 endemic mussels in this river alone. But when Williams reached into the mud and picked up a handful of clams, they were all exotic—*Corbicula fluminea*, an Asian clam that's been in the North America since the 1920s and is now established in more than 30 states. Known as the prosperity clam, it had been introduced by Asian immigrants; I had seen the same species along the shores of Lake Lanier.

One of the biologists ran her fingers across the dredge screen. It was Karen Herrington, the "meaningless bureaucrat," now several months pregnant. She handed me a clam that was deep black, about three inches long, with seven horizontal ridges that stretched across the shell like breakers in a storm. *Amblema neislerii,* a fat-threeridge. The interior of these animals, their soft anatomy, was a mystery. Even what fish species were essential to the development of their larvae was unknown.

During the drought, Williams and the Herringtons had helped move endangered mussels from above the receding shoreline down into the deeper channel of the river. They were going to have to do it again if the Corps continued to drop the water level. As we reboarded our boat, Williams recounted the heady days when he started work at the Fish and Wildlife Service in 1974. He had just been fired from a teaching position at the Mississippi State College for Women after participating in a lawsuit to prevent a boondoggle on the Tenn-Tom Waterway—that wasn't the way things were done down there, he'd been told. "It was probably the best thing that ever happened to me, because it got me out from behind the Magnolia Curtain."

Soon after Williams moved to DC, he started a package for the Red Hills salamander. "I'm a native of Alabama," Williams told me, "and I went to the University of Alabama so I know the Southeast pretty well. It was like, 'Goddamn, this salamander is just getting hammered by logging.'" He got some data from a local herpetologist. "I finally said, 'What the hell, I'm just going to list the salamander.' I didn't think too much about it. Well, the proposal comes out in the *Federal Register,* and the day it came out, I got a call from this guy in the Mobile *Press-Register.* He said, 'So I see your people have proposed to list the salamander. What do you think that's going to do to the area? Is there going to be some economic impact or kind of problem?' I played it down to the max, and the next day the headline read, 'Salamander to Stop Timber Harvest in Southeast.'" It became the first amphibian listed after the Act was passed.

We stopped to rake along a muddy bank on the Chipola, looking for the endangered slabshell, endemic to the river. After a couple of rakes, Williams turned up several, the color of a new penny once he rubbed the mud off. "Damned endangered mussels are everywhere," he joked.

"Mussels and snails have been the hardest-hit groups," Williams said. "One postal worker, a naturalist and conchologist, kept notes on the *Gyrotoma* snails in the sixties when I was a graduate student." Several species of this genus were found on the Coosa River in Alabama and Georgia. "No one in my department knew about it, and that was in the days when no environmental impact statements were done. Even if environmental statements were required, probably nothing would have been done. We'll never know. They completed the Weiss Dam and then Logan Martin Dam. As the mail-carrier collected snails in the rising water, he recorded in his notes, 'This is the last call for *Gyrotoma.*'" The entire genus has disappeared, along with dozens of other freshwater snails and unionids.

After we left Wewahitchka, Williams took me up to the Woodruff Dam, the nearest impoundment. For him, the days were for working, the late afternoons for dreaming. "I've been putting together 'Jim's Top Ten Dams for Removal.' On long trips like this, I daydream about what would I do. What kind of criteria would I use to evaluate which ones would come out first? Well, obviously the ones that have the least human benefits: money, flood-protection, power. At the same time, how many miles of river can I reclaim if I take it out? This one just happens to be at the top of my list. You have water being backed up into the lower Flint and the Chattahoochee, so if you take out one dam you're opening up the lower reaches of two rivers. You could recover some sturgeon populations and also Alabama shad and skipjack herring."

We were driving through the heart of the purple bankclimber's range. These oversize clams—Williams later showed me a shell that had the size and heft of a tomahawk—can live for 50 or 60 years, but their populations were well below critical levels. The dam was in the middle of prime bankclimber habitat but, to open the river for navigation, the Corps had blown up about ten of its shoals. A small population hung on at a limestone outcrop just south of the dam, the only remaining shoal where Gulf sturgeon spawned on the river. As we approached the dam, Williams pointed out a couple of small hills, Indian middens with the shells of thousands of freshwater mussels. The bankclimbers had been used for tools and food.

In the nineteenth and early twentieth centuries, unionids had been quite valuable. The discovery of a large pearl in the 1850s in an Ohio mussel started a pearl rush; hundreds of thousands of freshwater mussels were collected and destroyed. The pearls, soft and chalky, proved to be of little value; but, as the rush died down, a commercial button industry found a use for the iridescent nacre on the inside of the shell. By the 1910s, about two hundred button factories processed sixty thousand tons of shells per year, or about six hundred million mussels, the shell so valuable that harvesters used cannons to blow their competitors' boats out of the water. The mussel populations declined, and, with the development of plastic buttons, so did the harvest, only to be revived after World War II, when beads cut from mussel shells came to be used as the nucleus for cultured oyster pearls. (Nacre forms only around organic substances, not around sand.) Just about any cultured pearl is, at its core, a freshwater mussel.[9]

"These things are eventually going to fill up with sediment," Williams said, eyeing the dam. "It might be another 50 years or hundred years, but they're going to do it. How expensive is that going to be to deal with? It's not being used for navigation anymore. No one's operating the lock." The dam wasn't moving a lot of barges, except to reach the Choctawhatchee Nuclear Power Plant, but they could do that at high water, he added.

There was no one around as we drove up to the powerhouse in Williams's Subaru. There was a gleam in his eye. "There isn't anyone to stop you." We waited at the open gate. "You gotta start somewhere."

Most people will never see a purple bankclimber or a fat threeridge. In good times, they are completely covered by river water. You need a federal permit just to hold one. But there is another bivalve that's so celebrated in these parts, it has its own radio station.

I drove down to Cat Point, just east of the St. George Bridge, which crossed Apalachicola Bay on thick pylons, disappearing into the blue haze. Ten low-slung boats bobbed in view. On the closest one, two silhouettes in baseball caps set a small anchor. Against twenty-foot tongs positioned on the bow like a harpoon, one of them leaned as if he were splayed in chopsticks, then worried the rake through the sands. He pulled it onto the front deck and released a pile of oysters; a heap of them rose on the bow, where the second man weeded out what couldn't be sold. From the shore, the process looked slow and rhythmic, the ripples on the water sharp as ink.

Oysters are big business—1,200 licensed harvesters in the area and 25 packing houses, selling about 10 percent of the nation's harvest. Oysters on the Apalachicola are like blue crabs in the Chesapeake, crawfish in Louisiana, lobsters in Maine, or salmon in the Pacific Northwest: they're as much a part of the area's identity as the bay itself. They're also at risk—too little freshwater, and they can't survive.

Pickup trucks snuggled up against the tide, a few Sunday fishermen cast into the bay. A brown pelican glided over, the stealth pterodactyl landed on a pylon by the bridge. It was another success story for the Endangered Species Act. In the 1960s, the birds were in steep decline—largely as a result of DDT contamination. When they were listed in 1970, pelicans were rare in Texas, completely extirpated from Louisiana, and just holding on in Florida. Protections and reintroduction efforts—more than 1,200 pelicans were moved from Florida to Louisiana—helped the species come back from the brink. Delisted on the East Coast in 1985, it was declared completely recovered throughout its range in November 2009.

The oystermen and oysterwomen came ashore in boats mostly handmade—some wooden, low slung, with a cabin in the bow and a small outboard motor, others with the wheelhouse in the back. Some were modified Pro Lines, the fiberglass gunnels cut down so they rose just two feet above the water. The men looked as if they had spent a month at sea, thin, sun-beaten, some in skull-and-crossbones

T-shirts, others in muddied undershirts or tattered flannel, ciga-rettes dangling from parched lips. I caught up with them at Lynn's Quality Oysters, which packed and sold grouper, flounder, mullet, shrimp, oysters, and crabs, one of the few packinghouses left stand-ing as the supply of seafood dwindled. "It's easier on yourself if you keep it low," one fisherman told me, funneling oysters into burlap bags with a bottomless five-gallon bucket. Each 60-pound bag was tagged with the harvester's number, location of harvest, and the type and quality of shellfish. The burlap bags, piled on pallets until they were about shoulder high, would be shipped to Jacksonville or Tampa. The harvesters made $15 a sack.

One of the first to arrive was Earl Butler, in a chlorine-green baseball cap, plaid shirt, and jeans, his skin rippled from a life on the water. He was a man of few words at first—three, to be exact—when I asked him about the oyster harvest: "It's been better."

"Everything is out of balance," Marcus Stratton, an oyster buyer in wraparound sunglasses chimed in, being as bombastic and opin-ionated as the oystermen were tight-lipped. "You've got snapper where the bream's supposed to be. The shrimp's all up in the marshes. This is supposed to be a nursery for everything."

"All it is is Atlanta got too big for its britches," said Butler, who had stopped working construction after Katrina. "Since that storm hit in New Orleans, no one's building around here. Now there's nothing but oysters." And a few jellyballs: one of the fishermen told me that his father was harvesting cannonball jellyfish for the Asian market. The soft-bodied predators were the future of the oceans—and of seafood, if we weren't careful.

Apalachicola Bay was the oysterman's bank and heritage. The river fed blue crab, shrimp, grouper, drum, shad, and flounder. The seafood industry built around it was worth $134 million a year.[10] It supported the Sunday fishermen I had seen on the bridge, casting for an afternoon meal, and the large recreational boats. As we've seen from the spill of *Deepwater Horizon*, the loss of oysters hits more than the harvesters: it's a disaster for the people who refashion the burlap bags from sacks made for coffee beans; it affects the company that crushes the shells for chicken feed, literally; it hurts the shipping and the canning industries.[11] After the oil spill, the total loss of ecosystem services in the Mississippi River Delta could be in the billions.[12]

The Endangered Species Act isn't very good at protecting against rare catastrophes. The Fish and Wildlife Service and NOAA had largely ignored the low risk and high consequences of a blowout. When the oil was gushing, the happy ending to the stories of bringing back pelicans and sea turtles suddenly looked less certain.

Several studies have shown that protecting rivers for endangered species (such as bankclimbers and threeridges) provide benefits for communities downstream. The restoration of flow to the Rio Grande for the sake of the critically endangered silvery minnow would have a minor impact on local agriculture in central New Mexico, but farmers and ranchers downstream would get benefits of $400,000 a year; El Paso municipal and industrial users would reap more than $1 million a year.[13] Not bad for a three-inch algae-eater. As I was writing this book, opponents of the delta smelt in California grabbed endangered-species headlines—in what was another fight over water. In this case, saving the fish wouldn't help farmers who had turned the San Joaquin Valley into fields back when water had been plentiful and cheap; most economic analyses showed a high cost to farmers in holding the water back for the smelt, though salmon and fisheries would benefit from the restored flow. Not every case is win-win.

"It's nothing like it used to be," said Jeremy Register, another oyster buyer.

"We can start growing some if we get freshwater," Butler said. "They're holding our water back. Nature can always heal itself. But you gotta let it. And if you don't let the water back, it can't."

"Oyster will weigh different depending on what he's got to eat."

"There's a lot of little ones," Stratton added. "They don't have the freshwater to fatten them up."

A sign on the dock warned: "If youre number gets put on the board 1 time, you are fired. Your oysters are either to small or unculled."

The average haul was about ten bags—$150 for a Sunday, minus the gas. That was until Micheal J. Carmichael showed up, just before four. "He is the last one," Stratton said. "The last of the Mohicans." Carmichael had a slight paunch, a tobacco tin in his pocket, and deep lines that ran from the corner of his eyes to a blond and gray beard. He measured out 35 burlap bags. Water problems couldn't touch him.

"Y'all jam up," Stratton said. A boat that showed up late had him pointing to his watch. "I'm going to invest in Timex and buy them all watches."

Offshore a couple of fisherman cast a hand net, startling a shore-bird. Someone lit a joint. One fisherman said it was just as bad when too much freshwater was released at a time: "The oysters just pop open. They sound like Rice Krispies. Then they die." He held a straight face as he exhaled.

When I had arrived the night before, I had walked the docks in search of something to eat, but all of the restaurants had been closed. I ended up in a seaside bar with a beer and a couple of bags of chips, with some scientific papers I had brought along for company. The inability to predict how the day or the night would end drew me to such far-flung places, whether bars or wildlife sanctuaries—places where someone was as likely to wrap his arm around you and make you his nightlong drinking partner as to chase every Yankee in reach at least as far north as the Mason-Dixon line. A dark-haired woman pulled men from the pool table, wrapping her denim-clad legs around theirs and moving her butterfly tramp stamp to the music blaring. Few lepidopterans, listed or not, have had such attention.

I finished my beer and retreated to my cheap motel, spreading the unread papers on the bedside table. Science, at its most devilish, has the same air of the unknown. In the paper from the *Marine Ecology Progress Series* that I had been carrying around, a couple of scientists had lost half of their research equipment in two tropical storms that hit Turneffe Atoll in Belize in 2005. As they collected the pieces of their sediment traps and herbivore cages, marine ecologists Elise Granek and Ben Ruttenberg had been clever enough to record the damage. Never let a catastrophe go to waste. There had been anecdotal studies showing that storms became more dangerous as forests and dunes were removed from a coast, but Granek and Ruttenberg had the advantage of paired study sites: seven in all along a 30-kilometer stretch of coastline. Much of their equipment located behind mangrove forests had been left intact. Where trees had been cut down, the transects had been devastated.[14]

Healthy ecosystems can act as barriers to natural catastrophes. Mangroves once covered three-quarters of tropical and subtropical coastlines; less than half remains, and more than a third of these forests has been cleared in the past two decades. Following the tsunami in the Indian Ocean in December 2004, several researchers reported that mangroves left along coastlines were barriers to strong wave action, reducing property damage. Did this natural infrastructure save lives? When a super-cyclone struck Orissa, India, in 1999, there were fewer deaths in villages protected by large expanses of mangroves than in those where the salt-tolerant trees had been cleared to make way for rice paddies.[15]

In the United States, endangered species can help protect shorelines. Consider the diminutive beach mouse, three species of which live among the dunes of the Florida Panhandle, all of them endangered and likely never to be delisted. The Perdido Key beach mouse is nocturnal, monogamous, and completely dependent on the sea oats and bluestems that grow on the dunes of the Panhandle and eastern Alabama; there's just not enough habitat for this species to recover. But there's a bright side to protection: the dune plants essential for the mice also protect human seaside communities from tropical storms. The beach mouse, which disperses the seeds of these plants, likes high dunes. After Hurricane Ivan hit Florida and Alabama in 2004, there were some reports that houses near mice and protected dunes withstood the storm much better than those far from native plants.[16] The perverse incentives of coastal development that had the federal government insuring risky properties when no insurance company would touch them had been trumped by the Endangered Species Act.

As I left the oysterhouse, a big bird with a white bullet-shaped head flew across the road, then along the water's edge—and by sheer luck, when I pulled into a seaside neighborhood, I found its nest at the top of a loblolly snag. Fifty yards from the bay, on a research reserve, the eagle had chosen a beautiful location to raise three mottled juveniles. As it landed, its mate lifted up and disappeared from view, a black streak in an ochre sky. The eaglets gave out an occasional call, but didn't do much of anything for the hour I spent watching them—they were just the excuse I needed, after days on the road, to do nothing.

When I got back into the car, someone was walking on sunshine on Oyster Radio, "where the sun, sand, and sea meet." On the west side of the river sat the city of Apalachicola, across a concrete shame of a bridge that overshadowed the city, obstructing views of the water. In the old hotel in the center of town was a restaurant called Avenue Sea, where the oysters, just a buck apiece, were some of the finest I'd ever had: light and creamy, slightly salty, set off by a citrus and radish garnish. I had grown up on the firmer—and, yes, even saltier—northern oysters. These were almost too easy on the palate. I followed them with deepwater grouper, another species dependent on the Apalachicola. I remembered what Williams had told me on the river: "All the fish production is in those floodplain lakes. That's where they reproduce and have their nursery areas. People maybe ride up and down the river in boats, or cross the bridge here and there, and think, 'Well, there's still water in the channel so everybody's okay,' not realizing that the connectivity of our floodplain lakes to the main channel of the river is absolutely essential if there's going to be any productivity at all."

When the rains finally did arrive in northern Georgia in 2009, they fell in biblical proportions. Parts of Atlanta were flooded, at least nine people died, and Sonny Perdue declared a state of emergency in seventeen counties. But the threat of long-term drought remained.

David Carrier, the larger-than-life owner and chef of Avenue Sea, emerged from the kitchen and made the rounds, a green apron draped across his ample belly. "The oysters are like body-builders," he told me, flexing his long arms. "When the water is flowing, they bulk up, full of protein. They are delicious. But you can taste the salt water when the river is low. I think I'm used to it now."

18

At the edge of the pine forest, I stood at what had long been considered the only breeding pond of the Mississippi gopher frog. As we walked the drift fence, a foot-high corral made of roof flashing, Mike Sisson told me the frogs had moved into the ephemeral pond a couple of weeks earlier. Now we were looking for adults heading out of it.

Sisson retrieved one from one of the plastic bucket traps lining the fence. By the light of his headlamp, he showed me the marbled underbelly typical of the gopher frog, *Rana sevasa*. It had high ridges on its back and dark green spots on its arms. Sisson thought it looked like a female. He weighed it, measured its length from snout to vent, and scanned for a PIT tag—an electronic chip that would identify it, if it had been collected before. She had an orange dot on the inner thigh indicating that she had been hatched in the lab and raised in a cattle tank.

There having been little successful reproduction in the wild in the past nine years, it was no surprise that every frog that Sisson caught and set loose outside of the fence that night had already been tagged. The female looked at us with big black eyes as he estimated that she was at least five years old. As he was about to let her go, he noticed how tightly she was holding him.

"I'm going to change him to a male," Sisson said, as he erased the symbol in his Rite in the Rain notebook ("the best invention ever"). The frog's Popeye-style forearms showed that he'd had the full spring workout: nights of calling—metabolically as draining as running a marathon—followed, if he had been lucky, by amphiplexus, holding tight to the female and fertilizing her eggs as she released them.

We moved a leopard frog and a southern chorus frog to the other side of the fence. There was a white-footed mouse in one. I asked if Sisson wanted me to it take out.

"Sure, if you want to be a grizzly bear and grab it."

We came upon another gopher frog. "She's spent from her eggs," he said. "She's a skeleton." In his outstretched hand, she tried to cover her eyes but they were too prominent to hide. Sisson had found about 50 egg masses in the past year. "You couldn't ask for a better frog to work with."

I had heard gopher frogs in Florida after a late winter storm: the deep snore hugging the wet earth like fog, as if the world itself were asleep, but with a Looney Tunes edge, a mix that female gopher frogs found attractive. Sisson, a seasoned herpetologist, his camo hat flapping in the wind, certainly did: "It's the most beautiful sound in the world."

The wind picked up around us. Flashing kicked in the gusts. "I spend many, many hours on this damned drift fence. It's been my crucifix. I'll be coming out every night until midnight. Two years ago I had a ticket to a concert in New Orleans—and guess what happened? The first frog of the season showed up at the fence that night."

The next pail held only a short-tailed shrew—dead, in spite of Sisson's vigilance: shrews have such a high metabolism, they have to eat every two hours. If you didn't get to them soon enough, they died of starvation and exposure.

Sisson runs the trapline every night, not so much out of dedication to science, I suspect, but because cottonmouths hunt the fence. If he didn't move the frogs out of the buckets, the snakes could fill up on a critically endangered species. The flashing bucked up. "Watch your step," he quipped.

"It's an open-canopy pond; *sevosa* requires open ponds. We need to get a burn in through here." We piled back into his truck, and he lit an Ultra Light. (He was trying to quit.) "They filled the good ponds up, or they made them a fishpond, or for cattle-watering. In the forest, most of the temporary ponds are now wooded, probably because of Smokey the Bear." The longest-running public-service campaign recently turned 65. So famous he's trademarked, with his own zip code to handle all the mail he gets, he's an iconic patriot to some. Sisson blew smoke. "Smokey has ruined these woods."

For too long, fires were seen as something to avoid, rather than as an essential part of the forest ecosystem. And now there were master plans to build thousands of homes around this particular pond. "The frog wasn't listed when this was platted," he told me. "The big issue is fire. The weatherman puts up in big evil letters, every time the Forest Service does a burn, 'SMOG.'" Once the city is built, there will be plenty of resistance to controlled burns to protect mere frogs.

There were so many cracks in the windshield of Sisson's blue Toyota pickup that it looked ready to catch flies. He tried to resurrect a seatbelt from deep in the torn seat. It was hopeless. Sisson muttered about how all the federal funding went to research the obvious— everybody knows that gopher frogs don't like dense understory. "I hate the metric system," he continued, on a roll. "Try saying 'hectares' to someone." We drove past some trees snapped in half, some broken off at the sand—the legacy of Katrina. Young longleaf pines whipped around us. When I arrived at my hotel later that night, the TV meteorologist tracked a tornado cell just north of where we had been walking the fence. That gave me a thrill—but it highlighted the role that natural disturbance could play in extinction: a tornado hitting one breeding pond at the wrong time could kill all of the gopher frogs in it.

In 2003, there was an outbreak among the tadpoles of Glen's Pond. A newly discovered protozoan parasite, which resembled a pathogen common in oysters, had caused spore-like internal infections in the tadpoles. The parasite clustered in the liver, slowing the larvae down, swelling the abdomen, and causing them to swim erratically; infected frogs could easily be caught by hand. The outbreak killed just about every tadpole in the pond.[1] None reached adulthood in the wild.

A few hundred yards from the pond, Joe Pechmann, an associate professor at Western Carolina University, and his students maintained several cattle tanks filled with tadpoles removed from the pond. The larval frogs were red and black, granite pistachio nuts with two-inch tails, feeding on algae on the sides of blue poly tanks. They appeared to be thriving here, supplemented by a seasonal progression of plankton from another, presumably uninfected pond. Pechmann bleached his waders between ponds to avoid transmitting the disease. (Clorox is a donor to frog conservation.)

A bed of pine straw provided cover at the bottom of the cattle tank. Just below the surface, fairy shrimp did an odd backward dance. While Sisson and Pechmann worked the tanks, I walked most of the known range of the Mississippi gopher frog: longleaf pines, dirt roads, some recently cleared land. More trees on their sides, shorn off at the stump. At the sharp edge between the forest and the clear-cut, I wandered onto private land. Here was another legacy of Katrina: people were moving upland from the coast, and the state was four-laning smaller north-south routes.

Much of the gopher frog's world was about to become part of Traditions, a five-thousand-acre housing development. Rumor had it, one of the biologists told me, that the Fish and Wildlife Service had missed the potential threat because the development fell on the edge of a USGS map. By the time the Mississippi gopher frog's status came up for review, Traditions was stated as one of the reasons it needed protection.

It takes a few dedicated individuals to catch a species about to go extinct. Glen Johnson, who had seen other gopher-frog ponds disappear in the 1980s, discovered what's now called Glen's Pond in 1987. ("Rich Siegel named it that because he knew that it embarrassed me.") But if he hadn't gone out listening, it could have vanished as well.

"I followed that pond for years, on my own," Johnson said. In the first couple of years after he found the pond, there had been quite a few egg masses. "Two thousand four was the best year it looked like we were ever going to have. The tadpoles were really big—golf balls with tails. And then every one went belly up. And we thought, 'Well, it might be the end of them.'" Following gopher frogs as they entered the death spiral, going out on the rainiest, most miserable nights? "It's such a shitty job," he said, "I would only do it for free."

Johnson was eating peanut butter and jelly out of a squeeze pack, an MRE (government meal, ready to eat) left over from the response to Katrina two years earlier. He had retired from the Forest Service in 2004, and he and Sisson were looking for potential reintroduction sites for the frogs. There was plenty of open land nearby, the Forest Service owning much of it, but trying to find a place that could support a population in an artificial vernal pool was an engineering challenge—one that was new to the Service. "Before the recovery plan for red-cockaded woodpeckers went into effect," Johnson told me, "the Forest Service used to bulldoze their cavity trees."

Ornate chorus frogs had recently been lost in the area. "They're also in decline," Johnson continued, "but nobody seems to care." As we traveled the red sand roads of nearby Camp Shelby, we passed a sign threatening a $100,000 fine for capturing a gopher tortoise. Somebody was a good shot—about a dozen bullet holes had blasted through the pictured carapace. Shelby had the best longleaf pine forest in Mississippi—and the trees were still smoking from a prescribed burn. I could smell the char.

We climbed back into Sisson's truck. "For years, in Mississippi," he paused to hit his horn, "appraisers were devaluing property if a gopher tortoise was found on your land. So people thought, 'I better get rid of this turtle if it's going to cost me money.' It's just simply habitat. Habitat, habitat, habitat. If we kept their land, they could survive the bumps and scrapes, the ups and downs."

For gopher frogs, there were few good options on private lands. "The ideal pond would be three months," Sisson said. Even Glen's Pond needed water to extend the hydroperiod, delivered in tanker trucks in 2001 and pumped from a well in 2005. "Unfortunately, we're studying two-month ponds. The longer four-month ones were all turned into cattle ponds."

All except one. In 2004, thanks to a few resonant snores, Mike Sisson greatly expanded the known world of *Rana sevosa* when he had heard a deep snore in nearby Vancleave. One small pond was a huge discovery, doubling the number of known populations of the frog. Back then, Mike's Pond, as it's since become known, was relatively remote from human disturbance, there not being much but a cattle ranch nearby. Then came Katrina and the rush from New Orleans and the coast to resettle inland.

Johnson voiced what many biologists fear: "It would be neat to go back a hundred years—but they evolved the wrong strategy in modern times. These upland ponds are bisected by roads and opened to livestock. A single pond is just doomed. Eventually, you'll get a disease coming through or genetic diversity will decline. We should have done this years ago." Johnson took a drag on his cigarette. "Despite the fact that I think they're doomed, we kind of owe it to them to make an honest effort."

So we headed to his pond. "I came out one day," Sisson said, "and this land was bulldozed." Just a few months after the pond was discovered, a diesel-truck company had relocated from New Orleans. We parked in the company's gravel lot, a few feet from the water. "The Nature Conservancy couldn't buy land for more than it appraises, and the Fish and Wildlife Service has to find a dead endangered animal on the land to move on it. So Linda LaClaire," who was in charge of the gopher frog at Fish and Wildlife, "called M. C. Davis." Davis, Sisson explained, "made millions of dollars by developing the hell out of Florida, until he saw the light a couple of years ago." He bought the land.

That first year, Sisson had found 34 tadpoles. In 2005, there were egg masses, which could have made it, though nobody knows. Then in 2006, there was a devastating drought. In 2007, after Davis donated the land to The Nature Conservancy, prescribed burns were started. When I visited, there was still no documented evidence of successful reproduction, no evidence of young, sexually mature adults in the pond.

Unlike the disappearance of harlequin frogs in Latin America or gastric brooders in Australia, the loss of Mississippi gopher frogs wasn't unexpected. As with the red-cockaded woodpecker, they were devastated by the logging of the longleaf pine, by fire suppression, and by the conversion of timberland to plantations. The fragmentation of remaining habitats and the increase in mortality from road traffic probably brought the species to the brink—and then the disease swept in.

Here's what frogs are up against—climate change and the widespread chytrid fungus, things that set-asides can't protect them from. The couple of critically endangered Costa Rican frogs that were declared extinct in 2008 despite being protected by a national park are just two of dozens of species that have probably disappeared—the Global Amphibian Assessment lists 120 species possibly extinct, pending exhaustive surveys to verify their absence.

The frog-killing chytrid passes between amphibians through the skin. As the fungus infects the host, it turns the belly red; sloughed skin accumulates all over the body, followed by ulcers and hemorrhages. An infected frog turns lethargic; slow to seek shelter or flee from danger, it doesn't last long. There is some evidence that pesticides make amphibians more vulnerable to the pathogen, reducing the production of antimicrobial peptides crucial to fending it off. The global pet trade also endangers frogs because, thanks to air travel, the fungus comes into contact with even the most isolated populations. The bullfrog, native of North America, is such a popular pet in Japan that it has become invasive there—and it houses several strains of chytrid fungus. (The bullfrog itself resists infection but carries the pathogen.)[2]

Salamanders aren't doing much better. Several cloud-forest species in Mexico and Guatemala, abundant in the 1960s and 1970s, are all but gone now. *Pseudoeurycea smithi* once thrived in the mountains of Oaxaca, in narrow bands between 2,800 and 3,000 meters.

Flipping rocks along streams, researchers could uncover several hundred of the small salamanders in a single morning; in the past ten years, only a few have been found. As the climate has shifted, the salamanders haven't been able keep up with the temperatures: there's either nowhere to go once they hit the peaks, or the upstream habitats aren't quite right for them.[3]

Scientists at the University of Wisconsin have shown that excess nutrients—from farm runoff, fertilizers on suburban lawns, the paving of vast tracts of land—create conditions that allow some amphibian pathogens to thrive. The trematode parasite *Ribeiroia ondatrae* spends its early life in ramshorn snails; as algae bloom on ponds, snail populations explode. The tremadotes castrate the snail and emerge by the thousands, swarming on tadpoles, burrowing into the areas around their developing limbs and causing deformities. Amphibian survival plummets. The deformities may actually help the parasites by making the frogs an easier target for herons and other birds, their definitive hosts: birds that eat them pass trematode eggs to a new pond, spreading infection.[4]

It's more than just disease that endangers amphibians—it's our appetite for them. About a billion frogs are harvested from the wild each year, many from India, Bangladesh, and Southeast Asia; about ten thousand metric tons of frog legs are shipped internationally.[5] Most of them are imported into France, Belgium, and the United States. Besides being unsustainable, the trade may be another way of spreading the chytrid fungus around the world.[6]

At a seafood shack in nearby Woolmarket, Mississippi, I went to dinner with Sisson; Pechmann and his two grad students; Linda LaClaire, in charge of the gopher frog's recovery at the Fish and Wildlife Service; and Joe Mitchell, a consultant.

LaClaire, who wore a salamander brooch on her blazer, had disinvited me to a gopher frog meeting earlier that week. A successful lawsuit by the Center for Biological Diversity had forced the Service to act on designating critical habitat for the frog, and there were divisions among the herpetologists and between them and the Forest

Service. As far as I could tell, LaClaire didn't want an outsider reporting on the fray.

"It's hard working on a species like this that has one foot in the grave," she said. "It can get a bit contentious at times."

My order arrived, several crustaceans deep, on a plate big enough to cover a manhole; there were probably more crawfish on it than Mississippi gopher frogs left in the entire world. Bright red, delicious crawfish, mind you, but still in their shells. I was struggling. I might as well have had "Yankee" tattooed on my forehead. A woman from another table came up and demonstrated how to open the thing with a flick of her right thumb, as if it had a pop-off top. Break, twist, and suck. It was a sensual, experienced move. "But you might not be able to do that. Try it with two hands." She gently guided mine. I tried to keep up with her. It was hopeless.

She went back to her own plate at a table with her husband. I still had dozens of crawfish to make my way through, her technique remained unmastered. "She's one of the county supervisors," LaClaire whispered. They were all afraid that she had overheard their grumbling about local politics and development. Someone gave LaClaire a hard time for ordering shrimp. Eating dinner with conservation biologists was like walking through a minefield of ethical decisions: grasslands have been overgrazed by steer raised for beef, and all cattle emit greenhouse gases through enteric fermentation; the poop from industrially raised chickens poisons the Chesapeake; the Amazon has been slashed and burned for soy—and don't even mention seafood. To this bunch of herpetologists, the sin of ordering shrimp lay in the bycatch—young fish, and especially sea turtles, caught in the nets and discarded, dead or dying.

"Here we've got one pond on the edge of development and the other with part of its drainage basin in a semi-truck repair shop. The shop was moved to the site after the owner's previous shop was destroyed during Hurricane Katrina," LaClaire continued. As the climate changes, storms increase, and people—and other species—try to retreat from the coast, we'll no doubt see an increase in such conflicts. "Mike's Pond was going to be what Bruce Babbitt called a 'train wreck,' because an endangered species would be in direct conflict with development on private land" she said. "The landowner was planning on selling his land for a housing development. The consultant for the developer saw a meter stick and thought, 'This must be the feds, they're the only ones who would be using the metric system.'" The developer soon pulled out of the deal, and that, as Sisson had told me, is when Davis bought the land and donated it to The Nature Conservancy.

Like the red-cockaded woodpecker, the gopher frog was a holdover from a time not all that long ago when there had been vast expanses of longleaf pine and no fire suppression. But once the canopy closed around their ponds, temperatures dropped, slowing the growth and development of tadpoles, lowering the amount of dissolved oxygen available, and reducing the algal communities that they fed on. Removing trees from open-canopy ponds could help the frog recover. It could even protect the animals from the frog-killing fungus. "If a gopher frog senses it has chytrid," Pechmann said, "all it has to do is bask and the sun will get rid of it."

Prescribed burns are needed, but are they enough? In centuries past, fires that swept through vast tracts of pines were hot enough to take out the vegetation in the midst of ponds. Now the minor burns prescribed—where they are allowed—char the bark, but don't do much more than that. Sisson had pointed out a big pond in the midst of a farm, near where a road was due to go to four lanes. "It's just pork, man. It's just shit. This will be the death knell of the fire-managed ecosystem, right here."

On my way out of Gulfport, I passed a broad expanse of green. A perpetual-care park. Unfamiliar with the term, I thought of Johnson, Pechmann, and their colleagues moving egg masses and managing cattle tanks unceasingly into the future.[7] I slowed to check it out. It was a cemetery.

19

"There Is No Ivory Bill" is scrawled in black grease pencil across one of the signs at the Dagmar wildlife area in northeast Arkansas. A downy woodpecker called as I read about its history.

In the early 1940s, the Chicago Mill and Lumber Company began to cut down one of the last remaining stands of primeval bottomland in Louisiana—the Singer Tract, once owned by the sewing machine company—land that Jim Tanner, a Cornell graduate student who had studied the ivory-billed woodpecker as part of his dissertation, considered the last hope for saving the species. The National Audubon Society's attempt to buy the land before all the trees were felled had the support of the governor of Louisiana, but protests on behalf of the species only accelerated the speed of the saws. By 1944, most of the forest was gone. A lone female was spotted in a small strand of uncut timber, the last uncontroversial sighting of the bird.

As I drove to Lake Higgins, stopping to walk several trails in Dagmar, spiderwebs caught the first flakes of snow, suspending loose snowballs in the thicket. An otter crossed the road. By the time I encountered a pair of large woodpeckers, snow was falling so heavily, my binoculars fogged, and I never got a good glass on them. When I finally got a decent view, they were retreating into the bayou, letting out several high-pitched calls as they flew. Were they ivory bills? It would have to go down as the most exciting pileated woodpecker sighting I'd ever had in my life. Then again, every woodpecker in those particular woods made my stomach drop.

By the time I reached the lake, the park was covered in snow. It was almost a blizzard, with large flakes passing horizontally, never seeming to land, the end of the road a blank page. The snowy buttresses of cypress trees stretched up from the bayou like fish ribs sticking out from the muddy water. Six white-tailed deer bounded across the slushy road. A couple of guys dressed in thick camouflage came by in a red pickup—just wanted to come out and see it, they told me as the snow got even heavier.

"Elvis is everybody," came a voice on the radio. "Elvis is everywhere." "Elvis" had been the code name for the ivory bill, I would later discover, during the secret period before any findings were published. With no chance of a snowplow coming to my rescue should I get stuck in this Southern state, I reluctantly headed to Brinkley, population 3,940, rather than trace all the wild areas where the woodpecker had been sighted. Vast mixed flocks of birds, thousands upon thousands of starlings, grackles, and robins, morphed into scarves in the sky or darkened the fields, giving a Shakespearean voice to the barren trees. A grain elevator boasted, "The Heart of the Delta, Feeding the World." High notes chittered over the deep bellow of a train. It was 36 degrees on Cypress and Main, the downtown almost deserted, storefronts mostly dark rooms and "For Sale" signs.

At Gene's BBQ in Brinkley, the man at the door looked familiar—it was Gene DePriest, the owner, who I had seen driving the red pickup in the snow out at Dagmar. Decorated with paneled walls and peckerwoods—the local term for the bird—Gene's felt like ivory bill central. A huge poster in the restaurant celebrated the discovery of the bird in 2004. Birders were welcome here; DePriest told me that a group of twenty had stopped by earlier in the week. But when I visited all of the customers were locals: men in camouflage, eyeing the snow. I sat down by the window.

"That little girl bites just like a snapping turtle," I heard someone say. "I'd turn around and smack her in the mouth."

"You can't do that. You'll end up in the jailhouse."

"But she's a mean little rat."

The men agreed that it was time to go coon hunting: "If you have a long spell of cold, they have to come out."

The restaurant still sported the legend, "Home of the Ivory Bill Woodpecker," its menu, "We believe." I had my choice of an Ivory Bill burger (beef) or Ivory Bill salad (chicken). For the most part, the town and its inhabitants had taken up the bird's cause—a far cry from Papa Al's in Oregon's Mill City, which had served Endangered Logger cheeseburgers and Spotted Owl soup during the battles of the 1990s.

Before reaching Arkansas, I had called up Gene Sparling, the paddler who had started this whole thing when he spotted a large white-saddled bird on Bayou DeView in 2004. I asked if he would join me on the water one afternoon. He lived more than two hours away, outside of Hot Springs, and rarely traveled out to the Cache River. "I don't have much reason to," he told me. Wasn't he still looking for the ivory bill? "No. The ivory bill is the platinum bleached blonde," he told me, "who attracted all the attention. But there is a pretty farm girl behind her. The bird kind of helped open the world's eyes to what a treasure the Big Woods are."

Not everyone would like this comparison—farm girls, perhaps, platinum blondes, maybe even ivory bills, if they were still around to care—but he had a point. Since Congress had placed the emphasis on organisms—not on their habitats or ecosystems—endangered species often played a starring role in habitat protection. The northern spotted owl drew attention to the old-growth forests of the Pacific Northwest. The purple bankclimber and the fat threeridge sat above the fold of the *Wall Street Journal,* the poster mollusks for the Apalachicola. The snail darter was a three-inch David fighting for the Little Tennessee. All of these species and their legal rights garnered the attention of the public and much of the wrath.

"I do become frustrated at times," Sparling said, "that there's all this doing over the bird and relatively little to the glaring blatant truth before us all that this habitat needs to be protected and restored."

The platinum blonde—the Endangered Species Act—should inspire us to move from species-level protection to an ecological understanding of our place in the world: that elusive sense of place. Ultimately, it's the habitat, the ecological community, that's doing all the work. The farm girl is a healthy bay producing fresh oysters taken off Cat Point. It's Tupelo honey from the flooded forests. It's clear running water. It's wood ducks hunted in winter. It's pine straw at the base of a flowering tree on a highway rest stop in South Carolina. It's a cypress forest, covered in snow, the locals coming out to stare at the rare magic. It's a colony of bats over a cotton field, feasting on the pests. It's the most beautiful sound in the world: a Mississippi gopher frog calling on a stormy night in February—or the subtle inhalation of a humpback before it dives.

But Sparling's choice of the word "bleached" wasn't lost on me. *There Is No Ivory Bill*. Several ornithologists, including leading woodpecker biologist Jerome Jackson and bestselling birder David Sibley, had made their doubts about the discovery known in venues that were far more permanent than grease pencil. A few weeks after the article announcing the discovery in *Science* had come out, Sibley and colleagues expressed their considered opinion that the images were little more than poorly viewed pileated woodpeckers. Sparling seemed to be having doubts about the whole search. "The only confirmed sighting that I have any confidence in is that first time. And I spent the next period of years tearing that swamp up."

Allan Mueller, the man in charge of The Nature Conservancy's efforts to locate the ivory bill, agreed to take me out in the field. He suggested that we leave our destination open. "Maybe there will even be something hot going on then," he said. "Who knows?"

The trail had gone cold. Mueller had suggested a canoe trip through Bayou DeView, but seven inches of snow had given him pause. Without snowplows, we would have to wait for the roads to clear. Traffic had slowed to a crawl, cars and trucks skidding off the road. Waiting for Mueller, I watched Interstate 40 from my hotel room. The owner of the motel had changed its name to the Ivory Bill Inn after the discovery had been announced in 2005. The Ivory Bill Nest, a modest souvenir shop, sold T-shirts and woodpecker tchotchkes, and a local salon offered an ivory-bill haircut: a Mohawk, dyed red. Brinkley seemed poised for a boom.

Sparling later told me the town's original name was Lick Skillet. "Robert Brinkley was building a railroad from Memphis to Little Rock. They built it to the point of the present-day city and hit this big swamp. They said, 'Whoa, what do we do now?' They ran out of money, and they ran out of engineering skill to get across it. So the railroad workers all camped right there, while Mr. Brinkley went back and tried to get more financing and figure out how he was going to get across this vast uncharted swamp. Until the interstate came along, there wasn't really a lot to speak of for Brinkley."

As with so many towns in the rural United States, the skillet looked pretty well licked. Sparling called the region "a soybean, rice, milo sort of a culture." It wasn't exactly a ghost town, but it had become a city of starlings and sparrows. Wal-Mart was thriving, even in the midst of the storm—people needed cheap gloves and coats. McDonald's, hugging the interstate, easily outpaced Gene's BBQ. The birders had never flocked to Brinkley. Perhaps if it had stuck with "Lick Skillet," it could have retained more of its charm. Perhaps a B&B would have diverted a trickle from the river of motorists hurtling by at 70 miles an hour, knowing little of the elusive local celebrity. By the time I arrived in Brinkley, less than three years after the discovery of the bird had been made public, a new owner had taken over the Ivory Bill—it was now a Days Inn, the sunrise logo clearly visible from I-40. Robins bobbed on the yellowed lawn, and the major draw was HBO, not a bird once thought extinct; there were no Audubon caps or binoculars in the lobby while I was there.

Mueller showed up in a white pickup. When he asked if I wanted to take a snow check and go for a walk instead, I insisted I was still game for a paddle. On the way to the put-in site, I asked how the ivory bill's discovery had affected the local economy. After a brief bump, he told me, there hadn't been many visitors. He was hoping that the birds would turn up on private land, where someone could make a bundle from diehard listers: set up a blind and bring them in before the birds left in the morning—was a thousand dollars a person possible? I didn't think it was out of reach, considering how much people were willing to pay to see gorillas in Rwanda.

Mueller had retired from the Fish and Wildlife Service in 2006 and immediately stepped into his new position as head of the search for the ivory bill. He had more than twenty years of experience studying birds. "We won't see anything," he assured me, as he mounted a video camera—which he kept running the entire morning—on the canoe, hung his own still camera around his neck, and handed me a second camera for backup. "But I don't want to have to say that I wasn't ready." In the bottom of the canoe, he placed a wooden block, which could be struck to imitate the ivory bill's charismatic double knock.

The canoe edged through a thin layer of ice, past bare cypress trees covered in snow—it was like paddling through barracuda teeth. A pair of downy woodpeckers chased each other from cypress to tupelo. We passed a deep woodpecker scar on a tree. "The characteristic ivory bill would be just scaling the bark off. They do go deep, but so do lots of other woodpeckers. You don't see very much of that scaling."

Two wood ducks whistled through the cypress. The only other sound was snow melting off branches, and Mueller's soft narration: "This swamp was almost lost when the Army Corps of Engineers began dredging the Cache River in the 1970s." The Corps managed to dredge about six or seven miles before local protests closed the project down. Again, thanks to ducks: the ivory bill owes its existence—if it exists—to efforts to save the duck and duck-hunting. As the Cache was considered the most important mallard wintering site in the United States, a local dentist formed a coalition to save it, managing to stop the dredges just three miles from where we were paddling.

It was a bit like following the trail of the Civil War. "This is where Gene Sparling first saw the bird." A relatively open stretch of Bayou DeView. "This is where Bobby and Tim set up their camp. The genus of ivory-billed woodpecker is *Campephilus,* and when the Cornell people were at the Singer Tract in the 1930s, they named their campsite Camp Ephilus, a terrible joke, so naturally this spot became Camp Ephilus 2. An inside birder joke that isn't all that funny, but nevertheless. . . . And here's where the most famous four-second video in the world was shot."

The only moving image of an ivory bill in the past 60 years was taken with a video camera mounted on a canoe. Much of the frame is taken up by a knee, a right hand, and a yellow canoe paddle (it's an enlargement, roughly a sixteenth of the original shot) but in the distance, among the tupelo and cypress trees, there's a large bird flying away. John Fitzpatrick, head of the Cornell Lab of Ornithology, and colleagues filmed reenactments of the four-second flight with lifelike wooden models of both an ivory-billed and a pileated woodpecker. The ivory bill had a narrow black wing stripe, the pileated a large white underwing with dark trailing edges. The bird in the video was larger than a pileated. According to the researchers, the models left little doubt that the bird in the video was an ivory bill. To the untrained eye, discerning an ivory bill at that distance required a leap of considerable faith.

Mueller's description of his own sighting of an ivory bill, how it approached him, and then turned away before he could get his camera, left me with less doubt. "There was no noise. It was only like five seconds. I had a camera in my hand, and I couldn't react fast enough, the bird was flying toward me. I really didn't get anything until he turned around and flew away from me. I got a good look but a real quick look." He acknowledged that the evidence was scant, but it was hard to imagine a lifelong birder and local resident getting this wrong, no matter how much magical thinking was going on.

Just before the discovery was published, there was talk at the Fish and Wildlife Service of declaring the bird extinct and taking it off the list. Without revealing his source, Mueller called the regional office and told them perhaps they shouldn't waste their time. So they slowed the process down. "There is now a recovery plan for the ivory bill," Mueller said. "And the first priority is to find and document that sucker. So we're implementing the recovery plan right now."

"It must be tough to implement a recovery plan for an animal that . . ." I halted on the words.

"For an animal we know so little about?" Mueller helped. "I think it's the responsible thing for the Fish and Wildlife to do. There's enough evidence of its existence that they can't ignore it, even if people disagree with it."

Mueller's claim that five breeding colonies might be left was less convincing. "They persisted until now so there has to be at least a few of them." But if there really were several clusters, how had so many biologists missed them for so many months?

A woodpecker drummed in the distance. It was the usual mix, according to Mueller, " 'I'm here, this is mine, I'm here, I'd like to see you, especially if you're of the opposite sex.' The volume depends on the size of the woodpecker, but also what they're banging on. If they get a real resonant tree it makes a lot of noise."

We listened to a long series of beats. "The ivory bill doesn't do that," Mueller said. "He just does the double knock." Here on the bayou, I was learning about the ivory bill from what it was not.

"I say there are five pairs, but I don't have the slightest idea, really. Inbreeding's gotta be an issue. The area is spread out over half a million acres. Where are they? Can they find each other?"

We stopped for a Power Bar lunch. Mueller took out a wooden box and gave it a few double knocks with a thick dowel. "I'm here, I'd like to see you." No response. We got back into the canoe. The landscape was melting, snow falling all around us, on the camera, on our heads, floating by in the brown mud.

"I'm ready to lose all of it," said Mueller, "and let the leaves grow. Let's have spring. The wood ducks are already nesting, I'm sure." He paused. A clump of snow melted off a tupelo. "And so are the ivory bills."

In the year following his first sighting of the bird, Gene Sparling had put more than sixty-two thousand miles on his '96 Toyota pickup, making the 240-mile round trip between Hot Springs and Brinkley several times a week. "I spent much of that time sleeping two or three hours a night," he said, "getting up at three a.m. and driving to Brinkley, and getting home, trying to meet the school bus, or not meet the school bus, and get up and do it again the next day." The days were for birding, the nights for driving.

He invited me to his house outside Hot Springs. Sparling, who is in his fifties, with a graying beard and receding sandy-brown hair, was proud of his kayak and his knowledge of Bayou DeView. "I can sneak through that swamp on this boat. I can out-paddle any canoe, with a trolling motor or without. The strongest paddler you can throw at me—I can easily outpace them. They used to set up human mist nets to see the ivory bill—I can sneak through. I can fly. You wanna see the boat?" Sparling brought me into his garage, smiling proudly. "You can touch it if you want. I'm kidding, I'm kidding.

"I heard about a place called Bayou DeView," he told me, "and the legends said there were three-hundred-year-old cypress trees growing there. I wanted to see a three-hundred-year-old tree. The first time I paddled Bayou DeView, I was going through this beautiful little section, and I had put my paddle down and was leaning back in my kayak. I remember thinking, 'My God, I'm the luckiest guy on the planet.' I had this conscious thought that I wanted to put this away, and pull it back when I have trouble. I was frozen, having this magnificent daydream. Right after that, this sucker drops in."

And then, like Mueller, Sparling got that thousand-yard stare. "When I saw the bird, there was something about the black-and-white pattern, which is one way on the pileated, another on the ivory-billed.

"If you're a kid, and you're reading your field guide, and you look in there under the ivory bill, it'll say something like 'presumed extinct,' which captured millions of us kids. I'm hugely familiar with pileateds. This place is just covered in them. I got a nest right across from my house. Nests are everywhere. So I know a pileated, just instinctively.

"He was up above. There was weather moving in, and he was apparently above the canopy and started coming straight down the water channel. The stem count is so huge there that the channel is one of the few places to fly. My God, I thought, that's the biggest pileated I've ever seen. He was coming in and his wings were fully outspread.

"About that time he noticed me. I'm sure he'd never seen anything like a bright blue sea kayak with a half-human sticking out of the top of it. And so he kind of went, *whoa,* and flared his wings and dodged over to a tree, sat there for a few minutes, bounced to the back of the tree, did a woodpecker peek-a-boo, checking me out as he went up, and then flew off and continued on.

"After he landed on the tree, I noticed—and it was fortunate, because otherwise we wouldn't be here—that where the two wings met on the back was white. I knew that it wasn't a pileated woodpecker, and I knew, if it wasn't a pileated, that the only other species possible was an ivory-billed. But ivory-billeds are extinct. Don't be an idiot. They've been extinct my whole life. And I didn't realize their range extended that far north. To me, it was a thing of the Deep South. But, then again, I knew it wasn't a pileated. And if it wasn't a pileated, it had to be an ivory-billed. But ivory-billeds are extinct. Don't be stupid. And that's how it went. There's a groove in my brain now from where that thought was circulating."

He posted an intentionally vague reference to his sighting on the Arkansas Canoe Club's website in February 2004, but word soon got out. "I never even mentioned the term 'ivory-billed,' but it was like saying I saw Sasquatch. People took notice."

Sparling agreed to take Tim Gallagher, editor-in-chief of *Living Bird,* and Bobby Harrison, a nature photographer, to Bayou DeView. "They were making a God-awful racket," Sparling told me. "I was trying to encourage them to be quieter as we were getting into the area where I saw the bird." When he had gone ahead of them, he must have flushed the woodpecker. Sparling conjured it in the air again, with an almost mystical tone. A small group of birders, including the director of the Cornell Lab of Ornithology, soon learned of the discovery.

"The secret period was the golden period. We were able to just try to find the bird, explore, lay the groundwork for the conservation work that needed to be done." The bird was still called by its code name, Elvis. (Bayou DeView is only an hour from Memphis). One of Sparling's jobs was to keep 30 Yankees hidden in Cotton Plant, Arkansas, population 960. "It was admittedly an exciting adrenaline period." As they prepared to publish, the group decided not to reveal the exact locations that the bird had been sighted, to prevent an ivory rush. "We were going to say only 'In the Cache River Wildlife Refuge,'" Sparling said. "The announcement was rushed because we had a leak." Gallagher's book, already printed, gave all the locations where the bird had been seen. "It caused us great problems. We had already been to the authorities, saying, 'We're not going to reveal the exact locations—please don't close the refuge.'"

We drove over to Sparling's mother's house, a New England saltbox in the midst of Arkansas, where she was recovering from knee surgery, and he was looking in on her. His daughter showed up with his ex-wife, a short-haired brunette who greeted him rather draftily. I wondered if all those miles between Hot Springs and Brinkley, the missed buses and time away from home, had taken their toll.

"I'll cry like a baby when the photo of the ivory bill comes out," Sparling told me back in his kitchen. "I'll collapse on the ground. But it was time to leave it to the pros. I took my shot. I'm sorry I failed. I have been all over everywhere that I know to go. I couldn't find it again. Things happen. I learned through this project so, so, so, so many times that things happen as they should, and you have to just let them. I sort of dropped everything to go do this. It got to be time—well, that I needed to get my life back."

I reminded him that his sighting had helped preserve some of the habitat he adored.

"That really makes me feel good. Last check, I think we put something between sixteen thousand and eighteen thousand acres into conservation through a variety of means. That's what makes all the other turmoil and everything worthwhile. I remember being out there and thinking, gosh, if you could be responsible for saving forty acres of this, what a wonderful thing that would be. Hundreds of thousands of people drive by, within a stone's throw of Dagmar, the most pristine, mystical, enchanted swamp I have seen in my life. And nobody even knows it's there." Soon after our talk, Sparling stopped taking interviews and became—what else?—a woodworker, carving sculptures and bowls from native trees.

In a sense, the ivory bill has restored the idea of the inner frontier, a concept championed by Robert Kohler in his fascinating *All Creatures: Naturalists, Collectors, and Biodiversity, 1850–1950*. Early explorers in North America were confronted with a vast wilderness. But by the end of the Civil War, in the age of railroad booms and the homestead and timber acts (the largest giveaway of state lands in modern history), the American landscape became a mosaic of settlements and wild areas: Kohler's inner frontiers. In 1893, when he noticed that the westward movement of settlements was so scattered that there was no clear movement west, the historian Frederick Jackson Turner famously closed the American frontier. Settlers were just filling in the gaps. Naturalists started undertaking collecting expeditions and large-scale survey work as the infrastructure of roads and railways began to fall into place. During his faunal surveys in the early twentieth century, Joseph Grinnell relied on hotels as survey headquarters, using old stage roads as transects.[1]

Now, in the twenty-first century, was there really anywhere left for a twenty-inch black-and-white bird—the grail bird—to hide? The Big Woods feel big when you're in a canoe, almost lost in the cypress. But as Mueller pointed out, you're never all that far from a road. I could probably circle the entire known range—assuming, of course, that it is known—of the Arkansas ivory bill in my rental car in a day, without having to lose any sleep to the driving. As the snow melted, the cypress trees began to lose their white-toothed gleam. The interstate started to roar. Interior frontiers are too small to harbor something whose existence is as threatened as silence.

*   *   *

On my last day in Arkansas, I visited the Wattensaw Bayou, where Mueller had made his sighting. The ancient cypress, its dull brown catkins in bloom, cast shadowed slats across the water. An armadillo trailed through the leaf litter. The resurrection ferns were just starting to come back to life on the tupelo. Rarity and discovery opens up the inner frontier. Whether the ivory bill still has a chance or not, I think it likely that the Endangered Species Act could have helped prevent its disappearance in the first place. Allow me a bit of magical thinking. Back we go to the 1940s. There are still a few ivory bills remaining in the Singer Tract of northeastern Louisiana. A PhD student spends the next three years there, recording their movements. He concludes that only 22 remain, so the woods are preserved under the auspices of the Act and a search begins for other populations. Eventually, eggs are gathered, and adults swapped between isolated populations. The combination of habitat protection, captive breeding, and genetic rescues keeps the Lord God bird alive.

But could this have been done from just 22 individuals? Witness the 31 species of birds faced with imminent extinction that have been saved in recent years. Whooping cranes were down to 22 birds in the 1940s; a new refuge, captive breeding, and relocation efforts helped save them from extinction. The endangered Seychelles magpie-robin was saved by translocations to predator-free islands, the provision of nest boxes, and the control of the common myna and other invasive species. To save the Laysan duck, rabbits and invasive grasses were eradicated from one of the Hawaiian islands. The Mauritius parakeet, down to ten individuals in the 1980s, was saved from extinction through captive breeding and invasive species control; there are now about three hundred in the wild and a captive population held at a wildlife sanctuary on the island. In Ecuador, the pale-headed brush finch was rescued from cowbird parasitism and overgrazing by the protection of its scrubland; though its numbers are increasing, with just a square kilometer of habitat left, it remains critically endangered. All of these successes, according to conservation biologist Ana

Rodrigues, have made "a noticeable dent in the bleak scenario of global biodiversity loss."[2] All required focused and determined efforts.

But why let a species get to this state at all? It is always prudent, economically and ecologically, to save species before they get to the critical moment when captive breeding, translocations, and genetic rescues are required. It's already too late for the 1,175 animals and 747 plants on the Endangered Species List, or the 16,306 organisms on the IUCN's Red List, almost half of them flowering plants. We need a proactive approach to species protection, before they get to the conservationist's equivalent of the intensive care unit.

Just look at our ability to manage game. Besides the potential Indiana bat, I haven't noticed any endangered species crossing the view of my office window in Vermont. But a hundred years ago, many species that are now quite common might have found themselves on the list. By 1900, white-tailed deer had disappeared throughout the East. Trappers had extirpated beavers from Vermont decades earlier. The wild turkey population was barely hanging on.

Beavers were protected in 1910, and reintroduced from populations in New York and Maine in the 1920s and 1930s, as many farms were abandoned and forests regenerated. (The new brush and young trees were enough to allow the small white-tailed deer population to expand naturally, without new introductions.) Thirty-one turkeys from New York were released in the state in 1969. Thanks to restoration efforts and careful management, all of these species are now thriving, playing an important ecological and cultural role in the woods that surround my home. Even moose have returned to the northern forests. Conservation efforts can work, but they're just that—work. They're not magical thinking or set-asides restricted to the marginal lands of steep slopes and dramatic landscapes, such as the Grand Canyon or Yellowstone, as treasured as those places are.

There are steps that can be taken to steer us away from mass extinction, to approach the Holy Grail of conservation: zero extinction in our lifetime. We need to strengthen prohibitory regulations such as the Endangered Species Act. Although it may be decades before we can adequately assess its effectiveness, it is clear that protection works. If we see the glass as half full, most listed species improve or remain stable. Dozens more would have gone extinct without protection. Of course, the Act needs to be enforced and funded: just putting a species' name on a list won't take it out of danger. Some of this money should go to landowner incentive plans and endangered species banking so as to encourage win-win solutions. We need research on the suite of ecosystem services provided by rare and recovering species and their habitats. Without these studies, we'll continue to overestimate the costs of protection and ignore the many benefits.

What would life be like without the Endangered Species Act? It would be almost impossible to limit the impact of increasing human population size and economic growth. Consider the fate of the baiji, a freshwater dolphin of the Yangtze River, beset by a host of modern ailments: commercial fishing, vessel traffic, water projects, underwater explosions, and pollution. Just 50 years ago, there were about six thousand on the Yangtze; by the 1990s, estimates were down to about a hundred; only thirteen were seen during an intensive survey in 1997. In 2006, biologists on a 1,700-kilometer survey throughout its historical range who intended to capture all the remaining individuals didn't find a single one: no white humps on the river, no telltale clicks on the hydrophone. They did see, and hear, plenty of large ships—twenty thousand of them, one for every hundred meters of river surveyed.[3] Victim of economic growth and direct persecution, the baiji, the last representative of the family Lipotidae, may be the first recorded human-caused extinction of a cetacean species and the first large vertebrate known to have disappeared since the monk seal was last seen in the Caribbean over 50 years ago. International pressure for its conservation came too late.

The Yangtze giant softshell turtle may suffer a similar fate. As cities expand into farmlands and farmers move out to the margins, converting wetlands and forests to agriculture, the species has been reduced to just two known individuals, an 80-year-old female and a 100-year-old male. These freshwater turtles, the largest in the world, survive in two of China's zoos.[4] Land animals in China have fared no better. Elephants, gibbons, and snub-nosed monkeys, long ago hunted sustainably with dogs and spears, snares and blow-pipes, were reduced to tiny remnant populations in the twentieth century, thanks to firearms and large-scale habitat destruction. Rhinoceroses are already extinct there and throughout much of Southeast Asia. The South China subspecies of tiger has been elim-inated; all of the primates and most carnivores have suffered mas-sive losses. Half of the genera of the world's large mammals have gone extinct in the past fifty thousand years, many in the Western Hemisphere; Richard Corlett of the University of Hong Kong has noted that unless immediate action is taken, such mass extinction will have merely been postponed in tropical Asia.[5]

I sat on a fallen oak, my jacket zipped up against the chill of Southern winter. Bark beetles had scored rivulets in the fallen log. A leopard frog called from the slough. The woods were still. The snow was gone, except for the occasional white crescent wrapped around the base of a tree.

The days were for working, the nights for driving. I burrowed through the Arkansas night, the radio off. One of the strongest eco-logical texts I had read in recent years wasn't in *Science* or *Nature*.

They'd almost nothing left. . . . Out there was the gray beach with the slow combers rolling dull and leaden and the distant sound of it. . . . Out on the tidal flats lay a tanker half careened. Beyond that the ocean vast and cold and shifting heavily like a slowly heaving vat of slag and then the gray squall line of ash. He looked at the boy. He could see the disappointment in his face. I'm sorry it's not blue, he said. That's okay, said the boy.

An hour later they were sitting on the beach and staring out at the wall of smog across the horizon. They sat with their heels dug into the sand and watched the bleak sea wash up at their feet. Cold. Desolate. Birdless. . . . Along the shore of the cove below them windrows of small bones in the wrack. Further down the saltbleached ribcages of what may have been cattle. Gray salt rime on the rocks. The wind blew and dry seedpods scampered down the sands and stopped and then went on again.[6]

A world turned to ash, bereft of ecosystem services, the dustbowl of *The Grapes of Wrath* on a universal scale. It was *The Road* by Cormac McCarthy, one of the starkest and most moving visions of ecological bankruptcy—the complete loss of natural and social capital—I've ever read. One of the few bonds that remain is that between a dying father and his son. The healthiest environment they find is a fallout shelter, still filled with the dregs of the Anthropocene: "The faintly lit hatchway lay in the dark of the yard like a grave yawning at judgment day in some old apocalyptic painting." As in Bosch's final panel in *The Garden of Earthly Delights,* the burning road is the only escape for runaways and a conduit for their tormenters. A world beyond apocalyptic: a world that is post-nature.

"Conservation biologists don't practice conservation," Jeremy Jackson insisted over the phone. He and his wife, coral biologist Nancy Knowlton, have taught a graduate course in conservation at the Scripps Institute for Oceanography in San Diego for more than a decade. "We started to feel like all we were doing was writing, then refining obituaries for the planet. Is the work going to advance some policy, or is it some sort of necrophilia? Our students aren't interested in writing obituaries. They want to be practitioners."

So how do we get to zero extinction? How do we shepherd all species through the extinction crisis in a world with eight or nine billion people threatened by a possible world-changing temperature rise of 4°C? We've already laid out the technical tools, but how do we make extinction as unacceptable as slavery and child labor? Some might argue that it is natural—but there's nothing natural here. Paul Ehrlich at Stanford, Rob Pringle (his grad student at the time), John Avise at UC, Irvine, and I compiled nine steps that could take us down the road to success.[7] Many, such as the importance of ecological economics, are woven through this book. Here are more.

Biodiversity Parks. A carefully designed network of reserves on each continent and in every ocean is needed. Many traditional parks have been designed to protect geological formations or special features in the landscape: the volcanic caldera of Yellowstone, the Grand Canyon, or Mount Kilimanjaro. Others are on marginal lands, those that were left behind after people plowed through most of the fertile ones. A study of new reserves in Australia showed that they were typically gazetted on steep and infertile public lands, areas least in need of protection.[8] Such ad hoc reserves can be ineffective, occupying unproductive land and making the goal of protecting biodiversity expensive and less likely to succeed. A carefully designed network of reserves on each continent and in every ocean is needed. Rather than memorializing biocide or creating tiny reserves that are little more than perpetual-care parks, these protected areas should be part of an archipelago of reserves that protect endangered species and evolutionarily distinctive ecosystems. The task is not as daunting as it appears. By preserving and endowing just 25 biodiversity hotspots (less than 2 percent of the earth's land area), we can protect almost half of all vascular plant species and a third of mammal, bird, reptile, and amphibian species for $500 million a year—less than 1 percent of the funds the United States laid out for the Troubled Asset Relief Program to stabalize the banks. Of course, we shouldn't stop there, but such hotspots are a good place to begin.

In the United States, several hotspots of protection have emerged. Southern California, Hawaii, and Florida have areas with high degrees of endemism, where endangered organisms from different groups are found. San Diego County has numerous listed fish, mammals, and plants; Santa Cruz has arthropods, amphibians, reptiles, and plants. Since the first list was compiled in 1967, Hawaiian species have comprised approximately a quarter of all endangered and threatened organisms. By focusing our efforts in these areas, conservation can be efficient, with a large proportion of species protected on a relatively small amount of land.[9]

Biodiversity Trusts. One innovative way to establish and maintain protected areas is to create conservation trust funds. Modeled on US research universities, endowments provide sustained funding resilient to political changes and fluctuations in tourism. Conservation trust funds now exist in more than 40 countries. Some of the funds come from wealthy donors, visitors to national parks, and companies that bioprospect in protected areas or benefit from services such as clean water. Ideally, a board of trustees free from political entanglements administers the funds.

Diversity in Human Landscapes. Vast wilderness, protected from poaching and illegal timber harvest, is essential—but humans, at this point, are everywhere. Conserving diversity in and near human habitation has at least two advantages: more areas are open to protection, and a greater appreciation of wildlife is nurtured in people who can see it thriving close to home. Well-managed agricultural areas can help sustain many of the birds, mammals, and other organisms native to original forests. Even top carnivores such as pumas, jaguars, cheetahs, and wolves can survive on ranch and agricultural lands when owners manage their properties to allow for the coexistence of livestock and wildlife. Privately owned properties such as the Mpala Ranch in Kenya support lions, leopards, hyenas, and wild dogs in addition to healthy populations of domesticated and native grazers.

Our own daily actions should take wildlife into account—in rural, suburban, and even urban areas. The red-cockaded woodpecker should survive in Boiling Spring Lakes, not be banished to surrounding nature reserves. Here is where economists and ecologists can agree: ecosystem restoration is ecological engineering of the finest kind. It reassembles ecosystems that used to exist—longleaf pines or salt marshes—rather than creating new and untested combinations of species and habitats.[10]

Carefully designed Habitat Conservation Plans, strong Safe Harbor Agreements, and conservation banking can help conservationists and landowners manage working landscapes for target species and their ecosystems. Economic incentives, such as the US Department of Agriculture's Conservation Reserve Program, encourage biodiversity-friendly practices on private lands. Agricultural funds are often more dependable than those for endangered species. They also help populations of common species stay healthy rather than ending up in the ICU. Private NGOs such as The Nature Conservancy can play a crucial role in garnering support in communities and providing necessary funding to purchase and protect land.

Ecologically Reclaimed and Restored Habitats. We need to play conservation offense as well as defense. As with good medical care, let's focus on preventive conservation: protecting healthy wild areas, but also reclaiming the neglected and damaged ones. Let's focus on entire forests or grasslands, or on returning top predators—such as wolves—to areas where they have been absent for decades.

Moving wolves and other keystone species might actually help mitigate the effects of global warming. In an Alberta national park, researchers discovered that the presence of beavers increased wetland areas, regardless of changes in rainfall patterns. Beavers built small dams during droughts, blocking seepages and deepening ponds. They helped connect areas of open water, relatively rare where they were absent, by digging channels to draw water into their ponds—beaver irrigation. Reintroducing these rodents to areas where they once thrived could help engineer new, natural wetlands and limit the damage from future droughts.[11]

Firm Legislation. As the Endangered Species Act has made clear, legislation is essential. In the United States, lifting the Congressional cap on funding for new listings and critical-habitat decisions would go a long way in reducing the number of species that are languishing and sometimes declining while waiting to be fairly assessed. The appointment of Endangered Species Fellows, young or retired independent scientists with expertise in conservation and specific organisms who can work with Fish and Wildlife Service biologists during temporary assignments of a year or two, would speed up listing decisions and help depoliticize the process. In the absence of a full commitment from the federal government, litigation from groups such as the Center for Biological Diversity and WildEarth Guardians remains an essential tool in protecting species.

Smaller Ecological Footprints. Reducing consumption and stabilizing the human population, even humanely reducing it, will improve the lives of people and wildlife. As ecological economics has shown us, this requires a just distribution of resources. Or to paraphrase Leopold, you can't value nature without being assured of a good breakfast. These goals have been so often repeated that I almost didn't mention them—but without them, we have little chance of reversing the course of species' decline. It is nothing short of a reexamination of our lives and values.

Nobody ever said this was going to be easy.

In hindsight, I should have known better. But, after all the places I had visited in pursuit of endangered species, I wanted to end the book with one close to home—one I could see on foot. The Indiana bat was in the Class of '67—at risk, at least in part, because it hibernated in karst caves and abandoned mines, where it was subject to vandalism and disturbance. My home in Vermont was at the very northern edge of its range; in 2008, about three hundred Indiana bats were found in a dead elm three-and-a-half miles up the road. The females had used the tree for years to roost during the day and nurse their young.

I set out on the third of June. The sun was touching the Adirondacks, the cows already in the barn. A neighbor tended to her flowerbed. Asparagus was for sale at the Last Resort, an organic farm. These crops, the firmness of the road, and the thousand shades of green along Hogback Ridge were my calendar: we were just through mud season, summer barely here, and yet strawberries and dust would soon arrive, the trees deepening to a monochrome. Water spilled over a beaver dam.

The bats' roost was a tall, broken-down snag of an elm about ten yards from the road, not far from the Hidden Garden's B&B. The limbs had been sawn off, leaving odd-angled chimneys reaching to the sky. A birch tree, shining white, caught the breeze with the sound of well-maintained gears clicking in the wind. The sugar maple rippled, more of a whisper. Local woodsmen say they can identify a tree with their eyes closed, just by listening to wind in the leaves. For a moment, I believed them.

I stood by the elm. A great-crested flycatcher crossed the road. A male ruffed grouse flapped its wings—a deep-forest Harley—followed by a veery's eerie downward spiral. I paced back and forth on the road for a couple of hours. Except for the songbirds flying overhead, there was no sign of life until the mosquitoes started to bite. Not a single bat emerged.

By now, you must have heard the news: white-nose syndrome, a fungal pathogen known as *Geomyces destructans*, first recorded in New York in the winter of 2006, has spread rapidly through the Northeast. Little brown and Indiana bats have starved in their hibernacula. They dropped from their roof perches in caves the species had occupied since the retreat of the glaciers, ten thousand years ago. They died in hibernacula that were only a century old, ones mined for talc or marble. Bat carcasses, four or five deep, littered cave floors like the mast from chestnut trees (back before they also had been taken out by a fungus). Infected, having lost the fat reserves essential to surviving the winter, the bats flew off midseason in a last-ditch search for food. Some perished in mid-flight; others circled back to their caves in the dying light, clustering in a final effort to stay warm. They were seen clinging to frozen trees, making fallen angels in the snow. The populations in many hibernacula were reduced by more than 90 percent. Cave-dwelling bats could be extinct in Vermont by 2020.

When one dedicated researcher in the Adirondacks visited a cave in the winter of 2009, he found several thousand bats dead on the floor. *Geomyces* had colonized the skin around the muzzle, ears, and wings, replacing the hair follicles and sweat glands, invading the tissues beneath the epidermis. As he left, he noticed three individuals hanging on to the ceiling, their noses crusted with fungus. He couldn't bear to leave them, and snapped their necks before exiting the cave.

As far as I knew, I was the first to visit the roost tree that year. The sky was getting darker, the mosquitoes more aggressive. I didn't see a single living being leave the elm. Scott Darling, the state bat biologist, had estimated that the half million bats that overwintered in Vermont consumed about two billion insects a night. It was likely that at least a quarter of these bats had already disappeared. Would there be a rise in agricultural pests or mosquitos in areas affected by white-nose syndrome? If only I had some long-term data! At least these mosquitoes weren't malarial, but there was no guarantee against their carrying West Nile. The tree stood empty.

The loss of the bats was a loss of much more than pest control. "You can't help but have it tug at your heart," Darling said to one reporter, "tear at your soul." One of his colleagues told me that he had felt physically sick when he heard that I hadn't seen a single bat at Lewis Creek. I had hoped that I was doing something wrong.

Later in the season, I joined bat researcher Kristen Brisee at the site. "I remember walking the road last year," she said of the radio tagging she'd done there. "This was a bat highway." The mothers had left the tree, sometimes with their young on their backs, sometimes alone, foraged for a meal, and returned to rest and digest before flying off again. Brisee thought that the bats had found another home. I wasn't so optimistic.

That first night, I only saw one bat down the road, flying above a birch. It was chased off by a jay.

One night, I followed Brisee and several researchers up into the Green Mountains where they were mist-netting for bats. There were moose and ATV tracks on the rough road. (Someone whispered that with luck we might trap an ATV as bycatch.) We caught four relatively common little browns. Or should that be "once-common"? Even the old copper mines, once home to hundreds of thousands of bats, were largely empty now. All of the bats we caught had white staining on their wings, a sign that they were infected with *Geomyces*. They had made it through the winter, but would they last another?

The days were for working, the nights for driving. I had been up since five that morning, and left the researchers around two a.m. From the Middlebury woods to my house, it was all back roads, not a single gas station in sight where I could fill up—on caffeine. Maybe John Salyer, on his restless search for refuges, didn't have a two-year-old at home. I hooted myself awake on the road to Bristol and along

Monkton Ridge. By the time I turned onto State's Prison Hollow, two miles from home, I was jolted awake by the lash of wire and the crunch of cedar. I held tightly onto the wheel as a three-strand run of barbed wire sparked across metal and fiberglass, then did a 180 down a steep slope. Hot-pink sparks and the scent of smashed cedar lingered as my tires sank into a cattle pasture. I wondered if the car was totaled.

I called a tow truck and looked back up at the road. Somehow, I had just missed an elm, a flutelike 25-footer. Hadn't Dutch elm disease killed them all? The leaves looked healthy. Somehow it had escaped infection and—fortunately for both of us—the front end of my car. At least I hadn't hit a cow. Some chance events were near misses: once the tow truck got me back on the road, I was able to drive home and call the neighbors to fess up; the cows had already started wandering across the road. But for others, such as the spread of *G. destructans,* a chance event, if white-nose syndrome was a result of chance and not pesticides or pollution, can result in the near destruction of a species.

I wanted to believe that the bats would muddle through—but I couldn't ignore the record: chytrid fungus has taken out dozens of species of frogs in the past decade or so, spreading at rates of up to a hundred kilometers a year. The chestnuts were long gone from the region, along with most of the elms. Populations of Indiana bats had already dropped by more than half. Given the way microbes had taken out many once incredibly common species, it was hard to hold out much hope.

Was the epidemic natural? Or had the pathogen arrived on the dirty equipment of spelunkers who traveled from cave to cave, proud to look like West Virginia miners, gear soiled, as some have suggested? Cavers could have carried the spores from Europe, Asia, or who knows where. Then it spread from bat to bat. At this point, no one knows for sure. There was nothing natural here.

In a bat-lover's darkness, I walked home down three miles of mostly dirt road. I passed beneath a hemlock grove, its days, I regret to say, numbered, thanks to a tiny insect introduced from Asia. I crossed Lewis Creek under northern hardwoods—mostly sugar maples and paper birch. Through the lush, almost jasmine scent of black locust in bloom, I passed the meadow where we had first seen a fox when we had moved in five years ago. A man walked his little black dog. There was still a last shaft of light over the Adirondacks, a pregnant moon caught in the mist. Did I hear peepers? So late in the season? I humped over Hogback Ridge, the road steep enough to earn its name, State's Prison Hollow. Before we had bought our house, we had checked for prisons; there weren't any, nor had there ever been—before the road had been regraded, neighbors told us, it was said to be easier to break out of state prison than to get out of the hollow after a snow storm. The driver of the red pickup truck that sped around the bend cared nothing of this story, I suspected. By the time I reached home, I was enveloped in the calls of gray tree frogs—the sound of summer—and the baying of hounds up the way.

I walked through the darkened dining room, up stairs as steep and gray as a submarine's, across wide floorboards of yellowed pine. I sat at my desk. Out the small window, the barn was blood red. Windrows of hay lapped up against the copse. On my filing cabinet were a few shells and stones I had collected from the fallen dike of New Orleans's Lower Ninth Ward, two cherubs removed from the Old Met by my father before he had helped tear it down, a couple of graffitied shards from the Berlin Wall, and the shells of two rare clams that had died on the Apalachicola.

A month or so later, as I tried to coax our daughter to bed, we heard a rustling in the kitchen. There was a bat fluttering around the lights. It knocked a toy pirate off his perch above the window, then banked into the dining room.

"What's birdie doing?" Ngan asked, arms around my neck.

"It's just flying around," I told her. It was the first bat I had seen in weeks.

I closed the bedroom and bathroom doors, hoping it wouldn't head up the stairs where the cats would be waiting, one of them the tom that had once caught a little brown in the attic and left it at the foot of our bed.

"Moon!" Ngan yelled, pointing to the fuzzy light out the kitchen window.

As the bat glanced by our faces, I tried to react casually, but slipped back against the kitchen table.

"Okay, Daddy?"

This bat seemed larger than the others, and slower moving. If I opened the doors and left the room, the cats could escape—we didn't let them out at night—and the mosquitoes would get in. I didn't want to frighten my daughter, though I think I was the more scared—and the poor thing was flying into the walls. Ngan wasn't afraid of bats, not yet—as we had driven home from daycare, she had asked me to take her up the road to see them. I hadn't had the heart to tell her they were gone.

I opened the door to the natural world. We walked toward the bat.

Prologue

1. Associated Press, "Rare Woodpecker Sends a Town Running for Its Chain Saws," *The New York Times*, Sept. 24, 2006.
2. National Association of Home Builders, *Developer's Guide to Endangered Species Regulation* (Washington, DC: Home Builder Press, 1996), 109.
3. D. Lueck and J. A. Michael, "Preemptive Habitat Destruction under the Endangered Species Act," *The Journal of Law and Economics* 46 (2003): 27–60.
4. S. Reilly, "Sturgeon Protection Hasn't Endangered State Economy," Mobile (AL) *Press-Register*, Dec. 9, 2007.
5. C. D. Thomas et al., "Extinction Risk from Climate Change," *Nature* 427 (2004): 145–48.

1. In the Name of the Darter

1. Kenneth M. Murchison, *The Snail Darter Case: TVA versus the Endangered Species Act* (Lawrence: University of Kansas, 2007).
2. Ibid., 78.
3. David Etnier, *Proceedings of the Biological Society of Washington*, 88 (1976): 469–88.
4. Stanford Environmental Law Society, *The Endangered Species Act* (Palo Alto, CA: Stanford University Press, 2001). The courts have been divided on whether animals themselves can sue: in 1991, the Ninth Circuit ruled that the Hawaiian palila, an endangered finch, had the legal status to stand as a plaintiff, but a lower district court refused to allow

the endangered Hawaiian crow to be named as a plaintiff in a case against the Interior Department, in part because the environmental groups already had standing.

5. Sierra Club v. Morton, 405 U.S. 727 (1972).

6. Murchison, *Snail Darter*, 83.

7. Zygmunt J. B. Plater, "Reflected in a River: Agency Accountability and the TVA Tellico Dam Case," *Tennessee Law Review* 49 (1982): 747–87.

8. T. H. Ripley, "Letter to Ronald O. Skoog, Office of Endangered Species," Collection of James Williams, Gainesville, FL, 1975, 58.

9. Murchison, *Snail Darter*, 101.

10. Plater, "Reflected in a River."

11. Hill v. Tennessee Valley Authority, 549 F.2d 1064 (6th Cir. 1977).

2. The Class of '67

1. David Wilcove and Margaret McMillan, "The Class of '67," in *The Endangered Species Act at Thirty*, ed. D. D. Goble, J. M. Scott, and F. W. Davis (Washington, DC: Island Press, 2006), 1:45–50.

2. Endangered Species Preservation Act of 1966, Pub. L. No. 89–669, 80 Stat 926 (1966)(repealed in 1973).

3. Mark V. Barrow, Jr., *Nature's Ghosts: Confronting Extinction from the Age of Jefferson to the Age of Ecology* (Chicago: University of Chicago Press, 2009).

4. William T. Hornaday, *Our Vanishing Wild Life: Its Extermination and Preservation* (New York: Charles Scribner's Sons, 1913).

5. Ibid., 180.

6. Ibid., 177.

7. Aldo Leopold, *Game Management* (New York: Charles Scribner's Sons, 1933), 19.

8. Hornaday was a determined visionary, ahead of his time in compiling those early lists and in opposing the extermination of predators. But when he strayed away from birds, bears, and buffalo, Hornaday could be stubborn and wrongheaded. As director of the New York Zoological Park, he helped bring the African pygmy Ota Benga to the Bronx Zoo in 1906, displaying him in the monkey house. Local black leaders and clergymen protested. When Benga was released after two days, crowds hounded him on the zoo grounds. Hornaday considered releasing him

to the Brooklyn Howard Colored Orphan Asylum, but changed his mind when they couldn't come to terms. He insisted Benga be returned to Samuel Phillips Verner, an entrepreneur who had purchased him in a slave market in the Congo for a pound of salt and a bolt of cloth, intending him to be exhibited at the Saint Louis World's Fair. Verner liked to tell a different story: "I saved him from the pot and he saved me from the poisoned darts, and we have been good friends for a long time. I beg New York not to spoil him." Hornaday stubbornly refused to apologize. It didn't help that he singled out blacks in the South and Italian immigrants in the north in the destruction of songbirds. Benga later committed suicide in Lynchburg, Virginia, at the age of 32. Phillips Verner Bradford and Harvey Blume, *Ota Benga: The Pygmy in the Zoo* (New York: St Martin's Press, 1992), quote, p. 274.

9. Glover M. Allen, *Extinct and Vanishing Mammals of the Western Hemisphere with the Marine Species of All the Oceans* (Cambridge, MA: American Committee for International Wild Life Protection, 1942), vii.

10. D. B. Beard et al., *Fading Trails: The Story of Endangered American Wildlife* (New York: Macmillan, 1942), 86.

11. H. H. T. Jackson, "Conserving Endangered Wildlife Species," *Transactions of the Wisconsin Academy of Science, Arts, and Letters* 35 (1943): 61–89.

12. Barrow, *Nature's Ghosts,* 311.

13. Aldo Leopold, *For the Health of the Land: Previously Unpublished Essays and Other Writings,* ed. J. B. Callicott and E. T. Freyfogle (Washington, DC: Island Press, 2001), 264.

14. Rachel Carson, *Silent Spring* (Cambridge, MA: Houghton Mifflin, 1962), 33.

15. Ibid., 32.

16. Ibid., 85.

17. Alex Macgillivray, *Rachel Carson's Silent Spring* (New York: Barron's, 2004), 61.

18. "Rachel Carson Dies of Cancer; 'Silent Spring' Author Was 56," *New York Times,* April 15, 1964.

19. Carson, *Silent Spring,* 8.

20. 88 Cong. Rec. 16098–151 (1964).

21. I heard James Ellroy quote Chandler on the car radio while researching the book. As best I can tell, it's misquoted, but improves on the original.

3. Notes from the Vortex

1. Cited in Lawrence S. Earley, *Looking for Longleaf: The Fall and Rise of an American Forest* (Chapel Hill: University of North Carolina Press, 2004), 4.

2. Basil Hall, *Travels in North America in the Years 1827 and 1828* (Edinburgh: Robert Cadell, 1830), 3:252.

3. Earley, *Looking for Longleaf*, 195.

4. Ibid., 366.

5. Hall, *Travels*, 251.

6. Earley, *Looking for Longleaf*, 20.

7. E. V. Komarek, "Fire Ecology," *Proceedings of the Tall Timbers Fire Ecology Conference* 1 (1962): 95–107.

8. C. Vann Woodward, *Origins of the New South, 1877–1913* (Baton Rouge: Louisiana State University Press, 1951).

9. B. W. Wells and I. V. Shunk, "The Vegetation and Habitat Factors of the Coarser Sands of the North Carolina Coastal Plain: An Ecological Study," *Ecological Monographs* 1 (1931): 465–520.

10. David Ehrenfield, "Life in the Next Millennium: Who Will Be Left in the Earth's Community?" *The Last Extinction*, ed. Les Kaufman and K. Mallory (Cambridge, MA: MIT Press 1993), 195–214.

11. C. D. Huff et al., "Mobile Elements Reveal Small Population Size in the Ancient Ancestors of *Homo sapiens*," *Proceedings of the National Academy of Sciences* 107 (2010): 2147–52.

12. C. Richter et al., "Collapse of a New Living Species of Giant Clam in the Red Sea," *Current Biology* 18 (2008): 1349–54.

13. B. Van Valkenburgh and F. Hertel, "Tough Times at La Brea: Tooth Breakage in Large Carnivores of the Late Pleistocene," *Science* 261 (1993): 456–59.

14. G. C. Frison, "Experimental Use of Clovis Weaponry and Tools on African Elephants," *American Antiquity* 54 (1989): 766–84.

15. A. Lister and P. G. Bahn, *Mammoths: Giants of the Ice Age* (Berkeley: University of California Press, 2007).

16. Jacquelyn L. Gill et al., "Pleistocene Megafaunal Collapse, Novel Plant Communities, and Enhanced Fire Regimes in North America," *Science* 326 (2009): 1100–1103.

17. David Blockstein, "Letter to the Editor," *Science* 279 (1998): 1831.

18. J. A. Jackson and B. J. S. Jackson, "Extinction: The Passenger Pigeon, Last Hopes, Letting Go," *Wilson Journal of Ornithology* 119 (2007): 767–72.15.

19. J. Sapp, *Genesis: The Evolution of Biology* (Oxford: Oxford University Press, 2003).

20. Thomas Jefferson, *Notes on the State of Virginia* (Richmond, VA: J. W. Randolph, 1853), 55.

21. P. Del Tredici, "Against All Odds: Growing *Franklinia* in Boston," *Arnoldia* 63 (2005): 2–7.

22. Georges Cuvier, "A Discourse on the Revolutions of the Surface of the Globe," in *Evolution and Creationism: A Documentary and Reference Guide*, ed. Christian C. Young and Mark A. Largent (Westport, CT: Greenwood Press, 2007), 21.

23. R. M. May, J. H. Lawton, and N. E. Stork, "Assessing Extinction Rates," in *Extinction Rates,* ed. J. H. Lawton and R. M. May (Oxford: Oxford University Press, 1995), 1–24.

24. S. N. Stuart et al., "Status and Trends of Amphibian Declines and Extinctions Worldwide," *Science* 306 (2004): 1783–86.

25. E. La Marca et al., "Catastrophic Population Declines and Extinctions in Neotropical Harlequin Frogs (Bufonidae: *Atelopus*)," *Biotropica* 37 (2005): 190–201.

26. David S. Woodruff, "Declines of Biomes and Biotas and the Future of Evolution," *Proceedings of the National Academy of Sciences* 98 (2001): 5471–76.

27. J. Schipper et al., "The Status of the World's Land and Marine Mammals: Diversity, Threat, and Knowledge," *Science* 322 (2008): 225–30.

28. J. K. McKee et al., "Forecasting Global Biodiversity Threats Associated with Human Population Growth," *Biological Conservation* 115 (2003): 161–64.

29. Georgina Mace et al., "Biodiversity," in *Millennium Ecosystem Assessment: Ecosystems and Human Well-Being, Current State and Trends* (Washington DC: Island Press, 2005), 77–122.

30. Frederic Achard et al., "Determination of Deforestation Rates of the World's Humid Tropical Forests," *Science* 297 (2002): 999–1002.

31. D. Drew, *Man-Environment Processes* (London: Allen & Unwin, 1983).

32. C. J. Reading et al., "Are Snake Populations in Widespread Decline?" *Biology Letters* (2010): doi:10.1098/rsbl.2010.0373.

33. B. Sinervo et al., "Erosion of Lizard Diversity by Climate Change and Altered Thermal Niches," *Science* 328 (2010): 894–99.

34. A. Balmford, R. E. Green, and M. Jenkins, "Measuring the Changing State of Nature," *Trends in Ecology & Evolution* 18 (2003): 326–30.

35. Peter Kropotkin, *Mutual Aid: A Factor of Evolution* (New York: Cosimo, 2009), 9.

36. Earley, *Looking for Longleaf.*

37. D. C. Rudolph, H. Kyle, and R. N. Conner, "Red-Cockaded Woodpeckers vs. Rat Snakes: The Effectiveness of the Resin Barrier," *Wilson Bulletin* 102 (1990): 14–22.

38. A. Lopez-Sepulcre, K. Norris, and H. Kokko, "Reproductive Conflict Delays the Recovery of an Endangered Social Species," *Journal of Animal Ecology* 78 (2009): 219–25.

39. W. F. Fagan and E. E. Holmes, "Quantifying the Extinction Vortex," *Ecology Letters* 9 (2006): 51–60.

40. S. H. Strogatz, *Nonlinear Dynamics and Chaos* (Reading, MA: Addison-Wesley, 1994).

41. J. W. Kirchner and A. Weil, "Correlations in Fossil Extinction and Origination Rates through Geological Time," *Proceedings of the Royal Society B: Biological Sciences* 267 (2000): 1301–9.

42. Stephen M. Meyer, *End of the Wild* (Cambridge, MA: MIT Press, 2006), 4.

43. "The Ecological Footprint Atlas 2008," www.footprintnetwork.org.

44. National Wildlife Federation, *Standing Tall: How Restoring Longleaf Pine Can Help Prepare the Southeast for Global Warming* (Washington, DC: National Wildlife Federation, 2009).

45. W. W. Baker, *Observations on the Food Habits of the Red-Cockaded Woodpecker* (Folkston, GA: US Department of the Interior and Tall Timbers Research Station, 1971).

4. The Endangered Species Act

1. Charles C. Mann and Mark L. Plummer, *Noah's Choice: The Future of Endangered Species* (New York: Knopf, 1995), 156.

2. Endangered Species Protection Act of 1966, Pub. L. No. 89-669, 80 Stat 926 (1966) (repealed in 1973).

3. Mann and Plummer, *Noah's Choice,* 157.

4. Richard Nixon, Statement on Transmitting a Special Message to the Congress Outlining the 1972 Environmental Program, 50 Pub. Papers 183 (February 8, 1972).

5. Mann and Plummer, *Noah's Choice*.

6. 93 Cong. Rec. 21848 (1973).

7. Michael J. Bean, "The Endangered Species Act: Science, Policy, and Politics," *The Year in Ecology and Conservation Biology* 1162 (2009): 369–91.

8. Brian Czech, "The Capacity of the National Wildlife Refuge System to Conserve Threatened and Endangered Animal Species in the United States," *Conservation Biology* 19 (2005): 1246–53.

9. Robert E. Kohler, *All Creatures: Naturalists, Collectors, and Biodiversity* (Princeton, NJ: Princeton University Press, 2006).

10. Frederick Law Olmsted, "The Yosemite Valley and the Mariposa Big Trees: A Preliminary Report," *Landscape Architecture* 43 (1865): 12–25.

11. Roderick Nash, *The Rights of Nature: A History of Environmental Ethics* (Madison: University of Wisconsin Press, 1989).

12. Dale F. Lott, *American Bison: A Natural History* (Berkeley: University of California Press, 2003).

13. William T. Hornaday, *The Extermination of the American Bison* (Washington, DC: US National Museum, 1889), 388–89.

14. David S. Wilcove, *The Condor's Shadow: The Loss and Recovery of Wildlife in America* (New York: W. H. Freeman, 1999).

15. "Park Poachers and Their Ways," *Forest and Stream* 42 (1894): 444.

16. A. C. Isenberg, *The Destruction of the Bison: An Environmental History, 1750–1920* (Cambridge, UK: Cambridge University Press, 2000).

17. A. C. Bent, *Life Histories of North American Shore Birds* (New York: Dover, 1927).

18. D. B. Beard et al., *Fading Trails: The Story of Endangered American Wildlife* (New York: Macmillan, 1942).

19. Kohler, *All Creatures*.

20. "The Audubon Society," *Forest and Stream* 26 (1886): 41.

21. T. W. Cart, "The Lacey Act: America's First Nationwide Wildlife Statute," *Forest History* 17 (1973): 4–13.

22. Theodore Roosevelt's Seventh Annual Message to Congress (December 3, 1907).

23. John Muir, *A Thousand-Mile Walk to the Gulf* (1916; repr., New York: Mariner Books, 1998).

24. Nash, *Rights of Nature,* 41.

25. Ibid.

26. Theodore Roosevelt, *A Book-Lover's Holiday in the Open* (New York: Charles Scribner's Sons, 1916), 316.

27. W. Adams, *Green Development: Environment and Sustainability in the Third World* (New York: Routledge, 2001).

28. William T. Hornaday, *Our Vanishing Wildlife: Its Extermination and Preservation* (New York: Charles Scribner's Sons, 1913), 305.

29. K. Dorsey, *The Dawn of Conservation Diplomacy: U.S.-Canadian Wildlife Protection Treaties in the Progressive Era* (Seattle: University of Washington Press, 1998).

30. Thomas R. Vale, *The American Wilderness: Reflections on Nature Protection in the United States* (Charlottesville: University of Virginia Press, 2005).

31. John Lawrence, a self-styled literary farmer in England, looked on approvingly at the revolution across the Atlantic and, in 1796, penned a seven-hundred-page manifesto on the moral duties of man toward "the Brute Creation," proposing that "the Rights of Beasts be formally acknowledged by the state." Lawrence was mostly concerned with animal cruelty, but to Nash, it was a significant shift in the evolution of Western thought. Nash, *Rights of Nature.*

5. A Handy Handle

1. D. N. Greenwald, K. F. Suckling, and M. Taylor, "The Listing Record," in *The Endangered Species Act at Thirty,* ed. D. D. Goble, J. M. Scott, and F. W. Davis (Washington, DC: Island Press, 2006), 1:51–67.

2. Tennessee Valley Authority v. Hill, 437 U.S. 153 (1978), available at http://oyez.org/cases/1970-1979/1977/1977_76_1701.

3. Ibid.

4. Kenneth M. Murchison, *The Snail Darter Case: TVA versus the Endangered Species Act* (Lawrence: University of Kansas, 2007), 145–46.

5. US Dept. of the Interior, Endangered Species Comm. Hearing 26, statement of Charles Schultze, Chairman of the President's Council of Econ. Advisors (1979). Since its establishment, the committee has granted only two exemptions. In 1979, the Greyrocks Dam and the nearby $1.6 billion coal-fired power plant were relieved from ESA requirements. Though the flow of water was essential for whooping cranes that relied on the Platte River during spring and fall migrations, mitigation efforts appear to have

worked. In 1991, a vote of 5 to 2 (the minimum for passage) allowed the Bureau of Land Management to sell old-growth timber on the last habitat of the threatened northern spotted owl in the Pacific Northwest. But the ruling never went into effect; the Clinton administration withdrew the exemption, protecting ten million acres and reducing logging.

6. Zygmunt J. B. Plater, "Environmental Law in the Political Ecosystem—Coping with the Reality of Politics," *Pace Environmental Law Review* 17 (2000): 1201–65.

7. Comptroller-General of the United States, Report to the Congress: The TVA's Tellico Dam Project—Costs, Alternatives, and Benefits. EMD-77-58 (1977).

8. Tom Kenworthy, "The Protected and the Protector," *Trout* (Fall 2009): 19–25.

9. M. Burke and C. Milloy, "'Snakebitten' Andrus Refuses to Discuss Firing," *Washington Post,* Oct. 13, 1979.

10. F. X. Clines, "Watt Asks That Reagan Forgive 'Offensive' Remark about Panel," *New York Times,* September 23, 1983.

11. Greenwald et al., "Listing Record."

12. Zygmunt J. B. Plater, "Tiny Fish, Big Battle: 30 Years after TVA and the Snail Darter Clashed, the Case Still Echoes in Caselaw, Politics, and Popular Culture," *Tennessee Bar Journal* 44 (April 2008): 42.

13. Craig Welch, "The Spotted Owl's New Nemesis," *Smithsonian,* Jan. 2009.

6. Natural Capital

1. Margaret Schabas, *The Natural Origins of Economics* (Chicago: University of Chicago Press, 2006), Chapter 1 and cites therein.

2. Lisbet Koerner, *Linnaeus: Nature and Nation* (Cambridge, MA: Harvard University Press, 1999), 103.

3. Adam Smith, *The Theory of Moral Sentiments* (Oxford, UK: Clarendon Press, 1790/1976), 181.

4. Schabas, *Natural Origins,* 2.

5. John S. Mill, *On Nature* (Toronto: University of Toronto Press, 1874/1969), 10:380.

6. Nicholas Georgescu-Roegen, *The Entropy Law and the Economic Process* (Cambridge, MA: Harvard University Press, 1971), 11.

7. Herman Daly, "On a Road to Disaster," *New Scientist* (Oct. 18, 2008): 46–47.

8. R. Costanza et al., "The Value of Coastal Wetlands for Hurricane Protection," *AMBIO: A Journal of the Human Environment* 37 (2008): 241–48.

9. L. Emerton and E. Bos, *Value: Counting Ecosystems as Water Infrastructure* (Switzerland: IUCN, 2004).

10. M. L. Martinez et al., "The Coasts of Our World: Ecological, Economic, and Social Importance," *Ecological Economics* 63 (2007): 254–72.

11. Carol A. Kearns, David W. Inouye, and Nickolas M. Waser, "Endangered Mutualisms: The Conservation of Plant-Pollinator Interactions," *Annual Review of Ecology and Systematics* 29 (1998): 83–112.

12. R. W. Thorp and M. D. Shepherd, *Profile: Subgenus Bombus* (Portland, OR: Xerces Society, 2005).

13. M. Hickman, "New Life for the Ancient Black Honeybee," *Independent,* May 18, 2009.

14. Aldo Leopold, *A Sand County Almanac and Sketches Here and There* (New York: Oxford, 1949), 190.

15. S. Naeem, "Lessons from the Reverse Engineering of Nature," *Miller-McCune* 2 (2009): 56–71.

16. D. Tilman, D. Wedin, and J. Knops, "Productivity and Sustainability Influenced by Biodiversity in Grassland Ecosystems," *Nature* 379 (1996): 718–20.

17. J. F. Shogren, "Benefits and Costs," in *The Endangered Species Act at Thirty,* ed. J. M. Scott, D. D. Goble, and F. W. Davis (Washington, DC: Island Press, 2006), 2:181–89.

18. Jonathan Raban, "Losing the Owl, Saving the Forest," *New York Times,* June 25, 2010.

19. S. O'Connor et al. *Whale Watching Worldwide: Tourism, Numbers, Expenditures and Expanding Economic Benefits,* special report requested by the International Fund for Animal Welfare in Yarmouth, MA, prepared by Economists at Large, 2009.

20. Cutler J. Cleveland et al., "Economic Value of the Pest Control Service Provided by Brazilian Free-Tailed Bats in South-Central Texas," *Frontiers in Ecology and the Environment* 4 (2006): 238–43.

21. Andrew Balmford et al., "Economic Reasons for Conserving Wild Nature," *Science* 297 (2002): 950–53.

22. W. R. Freudenburg, L. J. Wilson, and D. J. O'Leary, "Forty Years of Spotted Owls? A Longitudinal Analysis of Logging Industry Job Losses," *Sociological Perspectives* 41 (1998): 1–26.

23. B. Martín-López, C. Montes, and J. Benayas, "Economic Valuation of Biodiversity Conservation: The Meaning of Numbers," *Conservation Biology* 22 (2008): 624–35.

24. I. J. Bateman and J. Mawby, "First Impressions Count: Interviewer Appearance and Information Effects in Stated Preference Studies," *Ecological Economics* 49 (2004): 47–55.

25. B. G. Norton, *Why Preserve Natural Variety?* (Princeton, NJ: Princeton University Press, 1987).

26. Mick Smith, *An Ethics of Place: Radical Ecology, Postmodernity, and Social Theory* (Albany: State University of New York Press, 2001).

27. S. M. Meyer, *End of the Wild* (Cambridge, MA: MIT Press, 2006).

28. Ibid.

29. The Journal of Henry D. Thoreau, entry for August 11, 1853, www .library.ucsb.edu/thoreau/writings_journals.html. As it turns out, even these chokeberries have a use; the dark fruits are rich in anthocyanins, antioxidants that could fight colon cancer, liver failure, and cardiovascular disease.

30. J. B. Callicott, "Explicit and Implicit Values," in *The Endangered Species Act at Thirty*, ed. J. M. Scott, D. D. Goble, and F. W. Davis (Washington, DC: Island Press, 2006), 2:36–48.

7. Magical Thinking

1. Chris Scott, *Endangered and Threatened Animals of Florida and Their Habitats* (Austin: University of Texas, 2004).

2. B. L. Burman, "The Glamour of the Everglades," *Reader's Digest* 72 (1956): 147–52.

3. D. C. Deitz and T. C. Hines, "Alligator Nesting in North-Central Florida," *Copeia* 1980 (1980): 249–58.

4. William Bartram, *Travels of William Bartram* (New York: Cosimo, 2007), 123.

5. Edmund Morris, *Theodore Rex* (New York: Random House, 2001), 519.

6. Nathaniel Reed and Dennis Drabelle, *The United States Fish and Wildlife Service* (Boulder, CO: Westview, 1984). Anne Vileisis, *Discovering the Unknown Landscape: A History of America's Wetlands* (Washington, DC: Island Press, 1999).

7. George Laycock, *The Sign of the Flying Goose: The Story of the National Wildlife Refuges* (Garden City, NY: Natural History Press, 1965), 9.
8. Thomas R. Vale, *The American Wilderness: Reflections on Nature Protection in the United States* (Charlottesville: University of Virginia Press, 2005).
9. Ibid.
10. E. Carver and J. Caudill, *The Economic Benefits to Local Communities of National Wildlife Refuge Visitation* (Washington, DC: USFWS, 2007).
11. My estimate, which is based on the number of acres protected for endangered species.
12. Brian Czech, "The Capacity of the National Wildlife Refuge System to Conserve Threatened and Endangered Animal Species in the United States," *Conservation Biology* 19 (2005): 1246–53.

8. Grand Experiments

1. Mark V. Barrow, Jr., *Nature's Ghosts: Confronting Extinction from the Age of Jefferson to the Age of Ecology* (Chicago: University of Chicago Press, 2009).
2. Ibid.
3. Faith McNulty, *The Whooping Crane: The Bird That Defies Extinction* (New York: E. P. Dutton, 1966).
4. M. H. Lytle, *The Gentle Subversive: Rachel Carson, Silent Spring, and the Rise of the Environmental Movement* (New York: Oxford University Press, 2007).
5. David H. Ellis et al., "Motorized Migrations: The Future or Mere Fantasy?" *Bioscience* 53 (2003): 260–64.
6. R. Southwick and T. Allen, "The 2006 Economic Benefits of Wildlife-Viewing Recreation in Florida" (2008), www.floridabirdingtrail.com.
7. Sahotra Sarkar, "Conservation Biology," in *The Stanford Encyclopedia of Philosophy*, ed. E. N. Zalta, www.plato.stanford.edu.
8. P. S. Alagona, "Biography of a 'Feathered Pig': The California Condor Conservation Controversy," *Journal of the History of Biology* 37 (2004): 557–83.
9. Ibid.
10. Charles C. Mann and Mark L. Plummer, "The Butterfly Problem," *Atlantic Monthly* 269 (1992): 47–70.

11. Alagona, "'Feathered Pig.'"

12. Lisa Gosselin, "On Human Intervention," *Audubon* (Jan./Feb. 2000): 6.

13. Dan Dagget, *Beyond the Rangeland Conflict: Toward a West That Works* (Flagstaff, AZ: Good Stewards Project, 1998), 73.

14. K. Alvarez, *Twilight of the Panther: Biology, Bureaucracy, and Failure in an Endangered Species Program* (Sarasota, FL: Myakka River Publishing, 1993), 343.

15. M. R. Matchett et al., "Enzootic Plague Reduces Black-Footed Ferret *(Mustela nigripes)* Survival in Montana," *Vector-Borne and Zoonotic Diseases* 10 (2010): 27–35.

16. R. Frankham, "Genetic Adaptation to Captivity in Species Conservation Programs," *Molecular Ecology* 17 (2008): 325–33.

17. H. Araki, B. Cooper, and M. S. Blouin, "Genetic Effects of Captive Breeding Cause a Rapid, Cumulative Fitness Decline in the Wild," *Science* 318 (2007): 100–103.

18. Frankham, "Genetic Adaptation."

9. The Panther's New Genes

1. Charles Bergman, *Wild Echoes: Encounters with the Most Endangered Animals in North America* (New York: McGraw-Hill, 1990), 94.

2. J. W. Pulliam, 56 Fed. Reg. 12950–52 (Mar. 28, 1991). Brian W. Bowen and Joe Roman, "Gaia's Handmaidens: The Orlog Model for Conservation Biology," *Conservation Biology* 19 (2005): 1037–43.

3. Warren E. Johnson et al., "Genetic Restoration of the Florida Panther," *Science* 329 (2010): 1641–45.

4. Carles Vilà et al., "Rescue of a Severely Bottlenecked Wolf *(Canis lupus)* Population by a Single Immigrant," *Proceedings of the Royal Society B: Biological Sciences* 270 (2003): 91–97.

5. Petter Wabakken et al., "The Recovery, Distribution, and Population Dynamics of Wolves on the Scandinavian Peninsula, 1978–1998," *Canadian Journal of Zoology* 79 (2001): 710–25.

6. P. L. Leberg and B. D. Firmin, "Role of Inbreeding Depression and Purging in Captive Breeding and Restoration Programmes," *Molecular Ecology* 17 (2008): 334–43.

7. H. B. Tordoff and P. T. Redig, "Role of Genetic Background in the Success of Reintroduced Peregrine Falcons," *Conservation Biology* 15 (2001): 528–32.

8. M. Culver et al., "Genomic Ancestry of the American Puma *(Puma concolor)*," *Journal of Heredity* 91 (2000): 186–97.

9. David S. Maehr, *The Florida Panther: Life and Death of a Vanishing Carnivore* (Washington, DC: Island Press, 1997), 54.

10. "Panther 2009 Death Toll One Short of Record," www.floridapanther.org.

11. Peter Kareiva et al., "Domesticated Nature: Shaping Landscapes and Ecosystems for Human Welfare," *Science* 316 (2007): 1866–69.

12. Peter M. Vitousek et al., "Human Appropriation of the Products of Photosynthesis," *BioScience* 36 (1986): 368–73.

13. Eric W. Sanderson et al., "The Human Footprint and the Last of the Wild," *BioScience* 52 (2002): 891–904.

14. Roderick F. Nash, *Wilderness and the American Mind*, 3rd ed. (New Haven: Yale University Press, 1982).

15. Charles Darwin, *The Variation of Animals and Plants under Domestication* (London: J. Murray, 1868), 242.

16. Michael J. Bean, "The Endangered Species Act: Science, Policy, and Politics," *The Year in Ecology and Conservation Biology* 1162 (2009): 369–91.

17. "The IUCN Red List of Threatened Species," www.iucnredlist.org (accessed Feb. 20, 2010).

18. J. M. Scott, D. D. Goble, and F. W. Davis, "By the Numbers," in *The Endangered Species Act at Thirty,* ed. D. D. Goble, J. M. Scott, and F. W. Davis (Washington, DC: Island Press, 2006), 1:16–35.

19. J. K. Miller et al., "The Endangered Species Act: Dollars and Sense?" *Bioscience* 52 (2002): 163–68.

20. D. Noah Greenwald, Kieran Suckling, and Martin Taylor, "The Listing Record," in *The Endangered Species Act at Thirty,* ed. D. D. Goble, J. M. Scott, and F. W. Davis (Washington, DC: Island Press, 2006), 1:51–67.

21. D. S. Wilcove and L. L. Master, "How Many Endangered Species Are There in the United States?" *Frontiers in Ecology and the Environment* 3 (2005): 414–20.

22. D. E. Davis, "Historical Significance of American Chestnut to Appalachian Culture and Ecology," in *Restoration of American Chestnut to Forest Land,* ed. K. Steiner and J. Carlson (Washington, DC: The North Carolina Arboretum, 2006). S. Freinkel, *American Chestnut: The Life, Death, and Rebirth of a Perfect Tree* (Berkeley: University of California Press, 2007).

23. A. P. Opler, "Insects of the American Chestnut: Possible Importance and Conservation Concern," *Proceedings of the American Chestnut*

Symposium, ed. W. L. MacDonald et al. (Morgantown: West Virginia University Press, 1978): 83–85.

24. S. A. Merkle et al., "Restoration of Threatened Species: A Noble Cause for Transgenic Trees," *Tree Genetics & Genomes* 3 (2007): 111–18.

25. Norman Myers and Andrew H. Knoll, "The Biotic Crisis and the Future of Evolution," *Proceedings of the National Academy of Sciences* 98 (2001): 5389–92.

10. Safe Harbor

1. D. Lueck and J. A. Michael, "Preemptive Habitat Destruction under the Endangered Species Act," *Journal of Law and Economics* 46 (2003): 27–60.

2. Donald J. Barry, "Amending the Endangered Species Act, the Ransom of Red Chief, and Other Related Topics," *Environmental Law* 21 (1991): 587–604.

3. Michael J. Bean, "The Endangered Species Act: Science, Policy, and Politics," *The Year in Ecology and Conservation Biology* 1162 (2009): 369–91.

4. Lawrence Buell, *Writing for an Endangered World: Literature, Culture, and Environment in the U.S. and Beyond* (Cambridge, MA: Harvard University Press, 2001).

5. "Andrew Mitchell: Forest Utility Meter Man," www.ecosystemmarket place.com.

6. N. Johnson, A. White, and D. Perrot-Maître, *Developing Markets for Water Services from Forests* (Washington, DC: Forest Trends, World Resources Institute, Katoomba Group, 2001).

7. City of New York Parks and Recreation, "Calculating Tree Benefits for New York City," (2007), www.nycgovparks.org.

8. R. S. Ulrich, "View through a Window May Influence Recovery from Surgery," *Science* 224 (1984): 420–21.

9. US Fish and Wildlife Service, "Army Recognizes Ralph Costa," www.fws .gov/southeast/news/2007/r07–094.html.

10. J. Fox and A. Nino-Murcia, "Status of Species Conservation Banking in the United States," *Conservation Biology* 19 (2005): 996–1007.

11. G. Heal, "Arbitrage and Options," in *The Endangered Species Act at Thirty*, ed. D. D. Goble, J. M. Scott, and F. W. Davis (Washington, DC: Island Press, 2006): 2:218–27.

11. Crying Wolves

1. E. Douglas Branch, *The Hunting of the Buffalo* (White Fish, MT: Kesinger, 2008), 222.
2. David A. Dary, *The Buffalo Book: The Full Saga of the American Mammal* (Athens, OH: Sage Books, 1989).
3. Rick McIntyre, ed., *War against Wolf: America's Campaign to Exterminate the Wolf* (Stillwater, MN: Voyageur Press, 1995), 15.
4. Horace M. Albright, "The National Park Service's Policy on Predatory Animals," *Journal of Mammalogy* 12 (1931): 185–86.
5. Horace M. Albright, "Correspondence to A. E. Demaray, Acting Director" (Washington, DC: National Archives, 1937).
6. Cited in Thomas R. Dunlap, *Nature and the English Diaspora* (Cambridge, UK: Cambridge University Press, 1999), 237.
7. Adolph Murie, *The Wolves of Mount McKinley* (Washington, DC: US Government Printing Office, 1944).
8. Curt Meine, *Aldo Leopold: His Life and Work* (Madison: University of Wisconsin Press, 1988), 181.
9. Susan L. Flader, *Thinking Like a Mountain: Aldo Leopold and the Evolution of an Ecological Attitude toward Deer, Wolves, and Forests* (Columbia: University of Missouri Press, 1974), 4. This story reminded me of Al Gore's evolution from voting to extirpate the snail darter on the Little Tennessee River to championing the fight against climate change. Perhaps he would be willing to discuss the trails he had followed before writing *Earth in the Balance* and *An Inconvenient Truth*? In researching the book, I tried to contact him a couple of times, but was told by his publicist that his schedule was full.
10. Aldo Leopold, *A Sand County Almanac and Sketches Here and There* (New York: Oxford University Press, 1949), 129.
11. John Goddard, "A Real Whopper," *Saturday Night* 111 (1996): 46–64.
12. Farley Mowat, *Never Cry Wolf: The Amazing True Story of Life Among Arctic Wolves* (New York: Dell, 1963).
13. Thomas R. Dunlap, *Saving America's Wildlife: Ecology and the American Mind, 1850–1990* (Princeton, NJ: Princeton University Press, 1991), 108.
14. Jan DeBlieu, *Meant to Be Wild: The Struggle to Save Endangered Species through Captive Breeding* (Golden, CO: Fulcrum, 1991), 161.

15. F. J. Rahel, C. J. Keleher, and J. L. Anderson, "Potential Habitat Loss and Population Fragmentation for Cold Water Fish in the North Platte River Drainage of the Rocky Mountains: Response to Climate Warming," *Limnology and Oceanography* 41 (1996): 1116–23.

16. C. C. Wilmers and W. M. Getz, "Gray Wolves as Climate Change Buffers in Yellowstone," *PLoS Biology* 3 (2005): e92.

17. Leopold, *Sand County Almanac,* 132.

18. John C. Morrison et al., "Persistence of Large Mammal Faunas as Indicators of Global Human Impacts," *Journal of Mammalogy* 88 (2007): 1363–80.

19. Ransom A. Myers and Boris Worm, "Rapid Worldwide Depletion of Predatory Fish Communities," *Nature* 423 (2003): 280–83.

20. James A. Estes and John F. Palmisano, "Sea Otters: Their Role in Structuring Nearshore Communities," *Science* 185 (1974): 1058–60.

21. J. K. Bump, R. O. Peterson, and J. A. Vucetich, "Wolves Modulate Soil Nutrient Heterogeneity and Foliar Nitrogen by Configuring the Distribution of Ungulate Carcasses," *Ecology* 90 (2009): 3159–67.

22. S. Koblmüller et al., "Origin and Status of the Great Lakes Wolf," *Molecular Ecology* 18 (2009): 2313–26.

23. T. M. Anderson et al., "Molecular and Evolutionary History of Melanism in North American Gray Wolves," *Science* 323 (2009): 1339–43.

24. J. W. Duffield, C. J. Neher, and D. A. Patterson, "Wolf Recovery in Yellowstone Park: Visitor Attitudes, Expenditures, and Economic Impacts," *Yellowstone Science* 16 (2008): 20–25.

25. US Fish and Wildlife Service, *The Reintroduction of Gray Wolves to Yellowstone National Park and Central Idaho: Final Environmental Impact Statement* (Helena, MT: USFWS, 1994).

26. S. Meyer, "Community Politics and Endangered Species Protection," in *Protecting Endangered Species in the United States: Biological Needs, Political Realities, Economic Choices,* ed. J. F. Shogren and J. Tschirhart (Cambridge, UK: Cambridge University Press, 2001), 138–65.

27. US Fish and Wildlife Service, "1.6 Million Acres Designated as Critical Habitat for California Red-Legged Frog," Press Release (2010), www.usfws.gov.

28. Zoological Society of London, EDGE of Existence (2010), www.edgeofexistence.org.

29. Shahid Naeem, Robin S. Waples, and Craig Mortiz, "Preserving Nature," in *The Endangered Species Act at Thirty,* ed. D. D. Goble, J. M. Scott, and F. W. Davis (Washington, DC: Island Press, 2006), 2:70–79.

30. Jorge Luis Borges, "John Wilkins' Analytical Language," in *Selected Non-Fictions* (New York: Penguin, 1999), 231.

31. Michel Foucault, *The Order of Things: An Archaeology of the Human Sciences* (London: Routledge Classics, 2002), xvi.

32. John C. Avise and Dale Mitchell, "Time to Standardize Taxonomies," *Systematic Biology* 56 (2007): 130–33.

33. O. A. Ryder, "Species Conservation and Systematics: The Dilemma of Subspecies," *Trends in Ecology and Evolution* 1 (1986): 9–10.

34. T. King et al., "Comprehensive Genetic Analyses Reveal Evolutionary Distinction of a Mouse *(Zapus hudsonius preblei)* Proposed for Delisting from the US Endangered Species Act," *Molecular Ecology* 15 (2006): 4331–59.

35. "Mouse Population's Future Split between Colorado, Wyoming," *Denver Business Journal,* Nov. 1, 2007.

36. "Building a Better Mousetrap," *Denver Westword News,* Jan. 20, 2005.

12. Skating over Thin Ice

1. Chris D. Thomas et al., "Extinction Risk from Climate Change," *Nature* 427 (2004): 145–48.

2. Michael J. Bean, "The Endangered Species Act: Science, Policy, and Politics," *The Year in Ecology and Conservation Biology* 1162 (2009): 369–91.

3. US Department of the Interior, "Secretary Announces Decision to Protect Polar Bears under Endangered Species Act," www.doi.gov.

4. Industrial Economics and Northern Economics, *Economic Analysis of Critical Habitat Designation for the Polar Bear in the United States,* prepared at the request of the US Fish and Wildlife Service (Cambridge, MA, 2010).

5. A. A. Gill, "The Maldivian Dilemma," *The Sunday Times,* Feb. 2, 2010.

6. Office of Inspector General, *Audit Report: The Endangered Species Program* (Washington, DC: USDOI, 1990).

7. D. Noah Greenwald, Kieran F. Suckling, and Martin Taylor, "The Listing Record," in *The Endangered Species Act at Thirty,* ed. D. D. Goble, J. M. Scott, and F. W. Davis (Washington, DC: Island Press, 2006), 1:51–67.

8. Conservation Council for Hawaii v. Lujan, 89-953 (D. Haw., 1998); California Native Plant Society v. Lujan, 91-0038 (E.D. Cal., 1991); Fund for Animals v. Lujan, 92-800, 962 F.2d 1391 (D. DC, 1992).

9. Greenwald et al., "The Listing Record."

10. M. L. Shaffer et al., "Proactive Habitat Conservation," *The Endangered Species Act at Thirty,* ed. D. D. Goble, J. M. Scott, and F. W. Davis (Washington, DC: Island Press, 2006), 1: 286–95.

11. NRDC v. US Dept. of the Interior, 113 F.3d 1121 (9th Cir. 1997). Karen E. Hodges and Jason Elder, "Critical Habitat Designation under the US Endangered Species Act: How Are Biological Criteria Used?" *Biological Conservation* 141 (2008): 2662–68.

12. Martin F. J. Taylor, Kieran F. Suckling, and Jeffrey J. Rachlinski, "The Effectiveness of the Endangered Species Act: A Quantitative Analysis," *Bioscience* 55 (2005): 360–67.

13. US General Accounting Office, *Endangered Species Act: The US Fish and Wildlife Service Has Incomplete Information about Effects on Listed Species from Section 7 Consultations* (Washington, DC: USGAO, 2009).

14. Juliet Eilperin, "Since '01, Guarding Species Is Harder," *Washington Post,* Mar. 23, 2008.

15. Office of Inspector General, *Report of Investigation: Julie MacDonald, Deputy Assistant Secretary, Fish, Wildlife and Parks* (Washington, DC: USDOI, 2007).

16. Office of Inspector General, *Investigative Report of the Endangered Species Act and the Conflict between Science and Policy Redacted* (Washington, DC: USDOI, 2008).

17. J. Becker and B. Gellman, "Leaving No Tracks," *Washington Post,* June 27, 2007.

18. US Department of the Interior and US Department of Commerce, "Interagency Cooperation under the Endangered Species Act," *Federal Register* 73 (2008): 47868–75.

19. Center for Biological Diversity & Maricopa Audubon Society v. Kempthorne, 2008 WL 659822 (D. Ariz. 2008).

20. J. D. Baker, C. L. Littnan, and D. W. Johnston, "Potential Effects of Sea Level Rise on the Terrestrial Habitats of Endangered and Endemic Megafauna in the Northwestern Hawaiian Islands," *Endangered Species Research* 3 (2007): 21–30.

21. Edward O. Wilson, *The Diversity of Life* (New York: W. W. Norton, 1992), 253.

22. J. W. Williams, S. T. Jackson, and J. E. Kutzbach, "Projected Distributions of Novel and Disappearing Climates by 2100 AD," *Proceedings of the National Academy of Sciences* 104 (2007): 5738–42.

23. Thomas et al., "Extinction Risk."

24. C. H. Sekercioglu et al., "Climate Change, Elevational Range Shifts, and Bird Extinctions," *Conservation Biology* 22 (2008): 140–49.

25. M. L. Rosenzweig, *Win-Win Ecology: How the Earth's Species Can Survive in the Midst of Human Enterprise* (New York: Oxford University Press, 2003).

13. Raising Whales

1. D. P. Gilmore, C. P. Da Costa, and D. P. F. Duarte, "Sloth Biology: An Update on Their Physiological Ecology, Behavior, and Role as Vectors of Arthropods and Arboviruses," *Brazilian Journal of Medical and Biological Research* 34 (2001): 9–25.

2. G. V Hilderbrand et al., "Role of Brown Bears *(Ursus arctos)* in the Flow of Marine Nitrogen into a Terrestrial Ecosystem," *Oecologia* 121 (1999): 546–50.

3. J. W. Kanwisher and S. H. Ridgway, "The Physiological Ecology of Whales and Porpoises," *Scientific American* 248 (1983): 110–20.

4. P. J. Corkeron, "Marine Mammals Influence on Ecosystem Processes Affecting Fisheries in the Barents Sea Is Trivial," *Biology Letters* 5 (2009): 204–6.

5. Herman Melville, *Moby-Dick* (New York: Norton Critical Edition, 2002), 121.

6. S. O'Connor et al., *Whale Watching Worldwide: Tourism Numbers, Expenditures, and Expanding Economic Benefits, a Special Report from the International Fund for Animal Welfare* (Yarmouth, MA, 2009).

7. Roger Payne, *Among Whales* (New York: Charles Scribner's Sons, 1995).

8. Jim Yuskavitch, *Conservation Pays: How Protecting Endangered and Threatened Species Makes Good Business Sense* (Washington, DC: Defenders of Wildlife, 2007).

9. Grace M. Johns et al., "Socioeconomic Study of Reefs in Southeast Florida 2000–2001," www.marineeconomics.noaa.gov/Reefs/02–01.pdf.

10. Kent E. Carpenter et al., "One-Third of Reef-Building Corals Face Elevated Extinction Risk from Climate Change and Local Impacts," *Science* 321 (2008): 560–63.

11. US Fish and Wildlife Service, "Wildlife Watching in the U.S.: The Economic Impacts on National and State Economies in 2006: Addendum to the 2006 National Survey of Fishing, Hunting, and Wildlife-Associated Recreation," (2008), www.wsfrprograms.fws.gov.

12. US Department of the Interior et al., "National Survey of Fishing, Hunting, and Wildlife-Associated Recreation," (2006), wsfrprograms.fws.gov.

13. C. H. Sekercioglu, "Impacts of Birdwatching on Human and Avian Communities," *Environmental Conservation* 29 (2002): 282–89.

14. J. E. S. Higham and M. Luck, *Marine Wildlife and Tourism Management: In Search of Scientific Approaches to Sustainability* (Wallingford, UK: CABI, 2008), 1–16.

15. David N. Wiley et al., "Effectiveness of Voluntary Conservation Agreements: Case Study of Endangered Whales and Commercial Whale Watching," *Conservation Biology* 22 (2008): 450–57.

16. I. A. Patterson et al., "Evidence for Infanticide in Bottlenose Dolphins: An Explanation for Violent Interactions with Harbour Porpoises?" *Proceedings of the Royal Society B: Biological Sciences* 265 (1998): 1167–70.

17. E. M. Scott et al., "Aggression in Bottlenose Dolphins: Evidence for Sexual Coercion, Male-Male Competition, and Female Tolerance through Analysis of Tooth-Rake Marks and Behaviour," *Behaviour* 142 (2005): 21–44.

18. Joe Roman and James J. McCarthy, "The Whale Pump: Marine Mammals Enhance Primary Productivity in a Coastal Basin," *PLoS ONE* 5 (2010): e13255.

14. Questing

1. H. Hoogstraal, "Changing Patterns of Tickborne Diseases in Modern Society," *Annual Review of Entomology* 26 (1981): 75–99.

2. Pamela Weintraub, *Cure Unknown: Inside the Lyme Epidemic* (New York: St. Martin's Press, 2008).

3. David Blockstein, "Letter to the Editor," *Science* 279 (1998): 1831.

4. J. S. Elkinton et al., "Interactions among Gypsy Moths, White-Footed Mice, and Acorns," *Ecology* 77 (1996): 2332–42.

5. Anne Gatewood Hoen et al., "Phylogeography of *Borrelia burgdorferi* in the Eastern United States Reflects Multiple Independent Lyme Disease Emergence Events," *Proceedings of the National Academy of Sciences* 106 (2009): 15013–18.

6. Sarah E. Perkins et al., "Localized Deer Absence Leads to Tick Amplification," *Ecology* 87 (2006): 1981–86.

7. Ronald Barrett et al., "Emerging and Re-Emerging Infectious Diseases: The Third Epidemiologic Transition," *Annual Review of Anthropology* 27 (1998): 247–71.

8. William H. McNeill, *Plagues and Peoples* (New York: Doubleday, 1998); A. J. McMichael, "Human Culture, Ecological Change, and Infectious Disease: Are We Experiencing History's Fourth Great Transition?" *Ecosystem Health* 7 (2001): 107–15.

9. Clark S. Larsen, "Biological Changes in Human Populations with Agriculture," *Annual Review of Anthropology* 24 (1995): 185–213.

10. N. D. Wolfe, C. P. Dunavan, and J. M. Diamond, "Origins of Major Human Infectious Diseases," *Nature* 447 (2007): 279–83.

11. Ibid.

12. D. A. Henderson, *Smallpox—the Death of a Disease: The Inside Story of Eradicating a Worldwide Killer,* with a foreword by Richard Preston (New York: Prometheus Books, 2009).

13. D. S. Wilkie and J. F. Carpenter, "Bushmeat Hunting in the Congo Basin: An Assessment of Impacts and Options for Mitigation," *Biodiversity and Conservation* 8 (1999): 927–55.

14. M. E. Woolhouse and S. Gowtage-Sequera, "Host Range and Emerging and Reemerging Pathogens," *Emerging Infectious Diseases* 11 (2005): 1842–47.

15. Asia Animals Foundation, "Species List" (2005), www.animalsasia .org.

16. Cahal Milmo, "Bush-Meat Trade Puts Britain at Risk of Ebola," *Independent,* June 23, 2006.

17. Dickson Despommier, *West Nile Story: A New Virus in the New World* (New York: Apple Tree Productions, 2001), 2.

18. Montira J. Pongsiri et al., "Biodiversity Loss Affects Global Disease Ecology," *Bioscience* 59 (2009): 945–54.

19. Stuart L. Pimm, "Lessons from a Kill," *Biodiversity and Conservation* 5 (1996): 1059–67.

20. A. M. Kilpatrick et al., "West Nile Virus Epidemics in North America Are Driven by Shifts in Mosquito Feeding Behavior," *PLoS Biology* 4 (2006): e82.

21. E. B. Hayes et al., "Epidemiology and Transmission Dynamics of West Nile Virus Disease," *Emerging Infectious Diseases* 11 (2005): 1167–73.

22. C. G. Hayes, "West Nile Virus: Uganda, 1937, to New York City, 1999," *Annals of the New York Academy of Sciences* 951 (2001): 25–37.

23. Vanessa O. Ezenwa et al., "Avian Diversity and West Nile Virus: Testing Associations between Biodiversity and Infectious Disease Risk," *Proceedings of the Royal Society B: Biological Sciences* 273 (2006): 109–17.

24. S. L. LaDeau, A. M. Kilpatrick, and P. P. Marra, "West Nile Virus Emergence and Large-Scale Declines of North American Bird Populations," *Nature* 447 (2007): 710–13.

25. J. P. Swaddle and S. E. Calos, "Increased Avian Diversity Is Associated with Lower Incidence of Human West Nile Infection: Observation of the Dilution Effect," *PLoS ONE* 3 (2008): e2488.

26. Laurie Garrett, *The Coming Plague: Newly Emerging Diseases in a World Out of Balance* (New York: Farrar, Straus and Giroux, 1994), 528–29.

27. Ibid.

28. James N. Mills, "Biodiversity Loss and Emerging Infectious Disease: An Example from the Rodent-Borne Hemorrhagic Fevers," *Biodiversity* 7 (2006): 9–17.

29. James N. Mills and J. E. Childs, "Ecologic Studies of Rodent Reservoirs: Their Relevance for Human Health," *Emerging Infectious Diseases* 4 (1998): 529–38.

30. L. J. Dizney and L. A. Ruedas, "Increased Host Species Diversity and Decreased Prevalence of Sin Nombre Virus," *Emerging Infectious Diseases* 15 (2009): 1012–18.

15. The Hundred Acre Wood

1. "Summary of Listed Species, Listed Populations, and Recovery Plans," ecos.fws.gov/tess_public/Boxscore.do (accessed Sept. 10, 2010).

2. "Federal and State Endangered Species Expenditures, Fiscal Year 2007," www.fws.gov/endangered/esa-library/index.html (accessed Oct. 12, 2009).

3. "IUCN Red List of Threatened Species: *Commidendrum Spurium*," www.iucnredlist.org.

4. Andrew P. Dobson et al., "Geographic Distribution of Endangered Species in the United States," *Science* 275 (1997): 550–53.

5. B. J. Cardinale et al., "Impacts of Plant Diversity on Biomass Production Increase through Time Because of Species Complementarity," *Proceedings of the National Academy of Sciences* 104 (2007): 18123–28.

6. P. J. Ferraro, C. McIntosh, and M. Ospina, "The Effectiveness of the US Endangered Species Act: An Econometric Analysis Using Matching Methods," *Journal of Environmental Economics and Management* 54 (2007): 245–61.

7. W. Cronon, *Uncommon Ground: Toward Reinventing Nature* (New York: Norton, 1995).

8. P. Lindemann-Matthies, "'Loveable' Mammals and 'Lifeless' Plants: How Children's Interest in Common Local Organisms Can Be Enhanced through Observation of Nature," *International Journal of Science Education* 27 (2005): 655–77.

9. Paul Bloom, "Natural Happiness," *New York Times Magazine*, April 19, 2009.

10. Adrian Franklin, *Animal Nation: The Story of Animals and Australia* (Sydney: University of New South Wales Press, 2006).

11. Scott McVay, "Prelude: 'A Siamese Connection with a Plurality of Other Mortals,'" in *The Biophilia Hypothesis*, ed. S. R. Kellert and E. O. Wilson (Washington, DC: Island Press, 1993), 3–19.

12. J. Pretty et al., "The Mental and Physical Health Outcomes of Green Exercise," *International Journal of Environmental Health Research* 15 (2005): 319–37.

13. Peter H. Kahn, Jr. et al., "A Plasma Display Window? The Shifting Baseline Problem in a Technologically Mediated Natural World," *Journal of Environmental Psychology* 28 (2008): 192–99.

14. Daniel Pauly, "Anecdotes and the Shifting Baseline Syndrome of Fisheries," *Trends in Ecology and Evolution* 10 (1995): 430.

15. R. Mitchell and F. Popham, "Effect of Exposure to Natural Environment on Health Inequalities: An Observational Population Study," *The Lancet* 372 (2008): 1655–60.

16. Richard A. Fuller et al., "Psychological Benefits of Greenspace Increase with Biodiversity," *Biology Letters* 3 (2007): 390–94.

17. Richard Louv, *Last Child in the Woods: Saving Our Children from Nature-Deficit Disorder* (Chapel Hill, NC: Algonquin Books, 2005).

18. Oliver R. W. Pergams and Patricia A. Zaradic, "Evidence for a Fundamental and Pervasive Shift Away from Nature-Based Recreation," *Proceedings of the National Academy of Sciences* 105 (2008): 2295–300.

19. Michael E. Soulé et al., "Ecological Effectiveness: Conservation Goals for Interactive Species," *Conservation Biology* 17 (2003): 1238–50.

20. David S. Wilcove, *The Condor's Shadow: The Loss and Recovery of Wildlife in America* (New York: W. H. Freeman, 1999).

21. Karen A. Bjorndal and Alan. B. Bolten, "From Ghosts to Key Species: Restoring Sea Turtle Populations to Fulfill Their Ecological Roles," *Marine Turtle Newsletter* 100 (2003): 16–21.

22. Joe Roman and Stephen R. Palumbi, "Whales before Whaling in the North Atlantic," *Science* 301 (2003): 508–10.

23. A. C. Isenberg, *The Destruction of the Bison: An Environmental History, 1750–1920* (Cambridge, UK: Cambridge University Press, 2001).

24. L. Chawla, "Ecstatic Places," *Children's Environments Quarterly* 7 (1990): 18–23.

25. Charlotte Braun-Fahrlander et al., "Environmental Exposure to Endotoxin and Its Relation to Asthma in School-Age Children," *New England Journal of Medicine* 347 (2002): 869–77.

26. C. A. Lowry et al., "Identification of an Immune-Responsive Mesolimbocortical Serotonergic System: Potential Role in Regulation of Emotional Behavior," *Neuroscience* 146 (2007): 756–72.

27. Y. Lai et al., "Commensal Bacteria Regulate Toll-Like Receptor 3-Dependent Inflammation after Skin Injury," *Nature Medicine* 15 (2009): 1377–82.

16. In Which We Upset the Ethnobotanists

1. G. J. Armelagos, "Take Two Beers and Call Me in 1,600 Years," *Natural History*, May 2000.

2. J. Goodman and V. Walsh, *The Story of Taxol: Science and Politics in the Making of an Anticancer Drug* (Cambridge, UK: Cambridge University Press, 2001); V. Walsh and J. Goodman, "The Billion Dollar Molecule: Taxol in Historical and Theoretical Perspective," *Clio Medica/The Wellcome Series in the History of Medicine* 66 (2002): 245–67.

3. F. A. von Hippel and W. von Hippel, "Sex, Drugs, and Animal Parts: Will Viagra Save Threatened Species?" *Environmental Conservation* 29 (2002): 277–81.

4. S. Proches et al., "Plant Diversity in the Human Diet: Weak Phylogenetic Signal Indicates Breadth," *BioScience* 58 (2008): 151–61.

5. Richard Wrangham and Nancy Lou Conklin-Brittain, "Cooking as a Biological Trait," *Comparative Biochemistry and Physiology* 136 (2003): 35–46.

17. Water Wars

1. Southeastern Federal Power Customers v. Geren, 06-5080 (D.C. Cir. 2008).

2. G. Bluestein and B. Evans, "Three-Year Countdown Begins for Atlanta's Water Future," Associated Press, July 21, 2009.

3. A. Carrns, "Atlanta Is Flexing Muscles in Its War on a Little Bivalve," *Wall Street Journal*, Oct. 26, 2007.

4. J. J. W. Rogers and P. G. Feiss, *People and the Earth* (Cambridge, UK: Cambridge University Press, 1998).

5. J. D. Williams et al., "Conservation Status of Freshwater Mussels of the United States and Canada," *Fisheries* 18 (1993): 6–22.

6. C. Lydeard et al., "The Global Decline of Nonmarine Mollusks," *Bioscience* 54 (2004): 321–30.

7. US Fish and Wildlife Service, *Biological Opinion on the U.S. Army Corps of Engineers, Mobile District, Revised Interim Plan for Jim Woodruff Dam and Associated Releases* (Panama City, FL: USFWS, 2008).

8. M. R. Darst and H. M Light, *Drier Forest Composition Associated with Hydrologic Change in the Apalachicola River Floodplain, Florida* (US Department of the Interior and US Geological Survey, 2008).

9. USGS Great Lakes Science Center, "Native Clams of the Great Lakes," www.glsc.usgs.gov.

10. Jenny Jarvie, "Atlanta Water Use Is Called Shortsighted," *Los Angeles Times*, November 4, 2007.

11. Dan Barry, "From a Gulf Oyster, a Domino Effect," *New York Times*, July 13, 2010.

12. R. Costanza et al., "The Perfect Spill: Solutions for Averting the Next *Deepwater Horizon*," *Solutions* 1 (Sep./Oct. 2010): 17–20.

13. F. A. Ward and J. F. Booker, "Economic Impacts of Instream Flow Protection for the Rio Grande Silvery Minnow in the Rio Grande Basin," *Reviews in Fisheries Science* 14 (2006): 187–202.

14. E. F. Granek and B. I. Ruttenberg, "Protective Capacity of Mangroves During Tropical Storms: A Case Study from 'Wilma' and 'Gamma' in Belize," *Marine Ecology Progress Series* 343 (2007): 101–5.

15. Saudamini Das and Jeffrey R. Vincent, "Mangroves Protected Villages and Reduced Death Toll During Indian Super Cyclone," *Proceedings of the National Academy of Sciences* 106 (2009): 7357–60.

16. Karen Voyles, "Mice to Nest at SFCC Zoo," *Gainesville Sun,* Jan. 14, 2008.

18. The Most Beautiful Sound

1. A. K. Davis et al., "Discovery of a Novel Alveolate Pathogen Affecting Southern Leopard Frogs in Georgia: Description of the Disease and Host Effects," *EcoHealth* 4 (2007): 310–17.

2. Goka Koichi et al., "Amphibian Chytridiomycosis in Japan: Distribution, Haplotypes and Possible Route of Entry into Japan," *Molecular Ecology* 18 (2009): 4757–74.

3. S. M. Rovito et al., "Dramatic Declines in Neotropical Salamander Populations Are an Important Part of the Global Amphibian Crisis," *Proceedings of the National Academy of Sciences* 106 (2009): 3231–36.

4. Pieter T. J. Johnson et al., "Aquatic Eutrophication Promotes Pathogenic Infection in Amphibians," *Proceedings of the National Academy of Sciences* 104 (2007): 15781–86.

5. I. Warkentin et al., "Eating Frogs to Extinction," *Conservation Biology* 23 (2009): 1056–59.

6. Brian Gratwicke et al., "Is the International Frog Legs Trade a Potential Vector for Deadly Amphibian Pathogens?" *Frontiers in Ecology and the Environment* 8 (2010): 438–42.

7. Their efforts would eventually pay off: in 2010, for the first time in more than a decade, 57 natural metamorphs, frogs born in the wild, were captured at the fence surrounding Glen's Pond and released on the other side.

19. The Platinum Blonde and the Farm Girl

1. Robert E. Kohler, *All Creatures: Naturalists, Collectors, and Biodiversity* (Princeton, NJ: Princeton University Press, 2006).

2. A. S. L. Rodrigues, "Are Global Conservation Efforts Successful?" *Science* 313 (2006): 1051–52.

3. S. T. Turvey et al., "First Human-Caused Extinction of a Cetacean Species?" *Biology Letters* 3 (2007): 537–40.

4. Jim Yardley, "Then There Were 2: Turtles' Fate Traces Threat to China's Species," *New York Times*, Dec. 5, 2007.

5. T. Corlett, "The Impact of Hunting on the Mammalian Fauna of Tropical Asian Forests," *Biotropica* 39 (2007): 292–303.

6. Cormac McCarthy, *The Road* (New York: Alfred A. Knopf, 2006), 215–16.

7. Joe Roman et al., "Facing Extinction: Nine Steps to Saving Biodiversity," *Solutions* 1 (Jan./Feb. 2010): 27–38.

8. R. L. Pressey et al., "Effectiveness of Protected Areas in North-Eastern New South Wales: Recent Trends in Six Measures," *Biological Conservation* 106 (2002): 57–69.

9. Andrew P. Dobson et al., "Geographic Distribution of Endangered Species in the United States," *Science* 275 (1997): 550–53.

10. A. Bradshaw, "Restoration of Mined Lands Using Natural Processes," *Ecological Engineering* 8 (1997): 255–69.

11. G. A. Hood and S. E. Bayley, "Beaver *(Castor canadensis)* Mitigate the Effects of Climate on the Area of Open Water in Boreal Wetlands in Western Canada," *Biological Conservation* 141 (2008): 556–67.

Acknowledgments

A huge thanks to Debora Greger, who assisted with the historical research, edited several drafts of the manuscript, and provided emotional and intellectual support throughout the project. If this were a scientific paper, she would be a coauthor. Ann Downer-Hazell encouraged me to start this book; Michael Fisher, editorial director at Harvard University Press, helped bring it across the finish line. Debbie Behler, editor of *Wildlife Conservation*, supported two trips to Florida. Caitlin Campbell was essential in checking many of the facts (though any mistakes are mine). Olivia Jacobsen put her years of text messaging to good use, helping with the transcriptions.

The Ucross Foundation gave me a residency in 2008, the Gund Institute for Ecological Economics provided an office in Vermont, and many of these ideas were spawned during a fellowship with the American Association for the Advancement of Science in Washington, DC. Montira Pongsiri and Iris Goodman led me down the path of ecosystem services, biodiversity, and human health. Bob Costanza welcomed me to the Gund Institute when I was just beginning the book.

Many people provided guidance and in-depth knowledge of their particular species. Here are a few that stand out: David Etnier, Pat Rakes, and Zyg Plater (snail darters); Pete Campbell, Ralph Costa, John Ellis, Ryan Elting, Jim Gray, John Hammond, Julian Johnson, Marty Kesmodel, David Lewis, Patty Matteson, Dan Ryan, Lea Anne Werder (red-cockaded woodpeckers); David Ellis, Lara Fondow, Steve Nesbitt, and Kathy O'Malley (whooping cranes); Michael Lockhart and Dean Biggens (black-footed ferrets); Dennis Giardina, Darrell Land, Roy McBride, and Dave Maehr (Florida panthers); Ed Bangs, Laurie Lyman, Rick McIntyre, and Doug Smith (wolves); Rob Ramey (Preble's meadow jumping mouse); Ari Friedlander, Pat Halpin, Elliott Hazen, Jim McCarthy, John Nevins, Cara Pekarcik,

Mason Weinrich, Dave Wiley, Jeremy Winn, Becky Woodward, and the crew and research staff aboard the *Nancy Foster* (humpbacks); David Newman, Kelly Oggenfuss, and Rick Ostfeld (biodiversity and disease); Jim Williams and Steve Herrington (purple bankclimber and fat threeridge); Glen Johnson, Linda LaClaire, Joe Pechmann, and Mike Sisson (gopher frogs); Kristen Brisee and Ryan Smith (Vermont bats); Alan Mueller and Gene Sparling (ivory bills); and many others who appear throughout the pages of this book. I also thank Michael Bean, Ken Dodd, Dale Goble, Noah Greenwald, Lee Talbot, and David Wilcove (Endangered Species Act).

Several people read sections of the book and offered corrections, both scientific and stylistic. Thanks to Don Barry, Ralph Costa, Laura Farrell, Michael Lockhart, Rick McIntyre, Zyg Plater, Frank Zelko, and three anonymous readers.

I am deeply grateful to my wife and daughter, who listened to my ideas, gave me some new ones, and held down the house while I was in the field. My family has been enormously generous and supportive: thank you Mom, Dad, Marie, and Joe and Theresa Sweeney.

And, finally, a heartfelt thank-you to all of the researchers, managers, and volunteers who dedicate their time and energy to protecting, restoring, and monitoring both the rare species and the common ones. Keep up the good work on the front lines.

Index